ETHICS FOR SCIENTIFIC RESEARCHERS

(Second Edition)

ETHICS FOR SCIENTIFIC RESEARCHERS

By

CHARLES E. REAGAN, Ph.D.
Kansas State University
Manhattan, Kansas

CHARLES C THOMAS • PUBLISHER
Springfield • Illinois • U.S.A.

Published and Distributed Throughout the World by
CHARLES C THOMAS • PUBLISHER
BANNERSTONE HOUSE
301-327 East Lawrence Avenue, Springfield, Illinois, U.S.A.
NATCHEZ PLANTATION HOUSE
735 North Atlantic Boulevard, Fort Lauderdale, Florida, U.S.A.

With THOMAS BOOKS *careful attention is given to all details of
manufacturing and design. It is the Publisher's desire to present books
that are satisfactory as to their physical qualities and artistic possibilities
and appropriate for their particular use.* THOMAS BOOKS *will be true
to those laws of quality that assure a good name and good will.*

Printed in the United States of America
EE-11

PREFACE

By 1945 it became undeniable that whether we like it or not, science and scientific research are not ethically neutral. The position, held for so many decades, that science is the search for pure knowledge and that this search is outside the scope of ethics, is no longer tenable. No one denies that the acquisition of knowledge is an intrinsic good, but that good is now seen to be but one among many. Furthermore, the enterprise of acquiring scientific knowledge has grown so large and so complex that it affects everyone, whether within or without the enterprise itself.

Some of the ethical problems which can arise in the course of scientific research and technological advancement have been made famous by the popular press. As examples, one might cite the current discussions of organ transplants and of air and water pollution. Most scientists can point to many ethical issues involved in the research presently being conducted in their fields. Recognition of these problems—though long in coming—is no longer of prime importance.

"How are we going to handle the problems?" and "What are we going to do about them?" are the currently pressing questions. This book is a first attempt to answer these questions. To answer an ethical question in any situation requires both an adequate grounding in normative ethics and the relevant factual information. Since this book is directed toward students in science curricula and practicing scientists, the latter requirement is presupposed. Thus, the sole aim of this effort is to supply an elementary grounding in ethics. However, it is not enough to merely know some of the fundamentals of ethics. One must be able to apply them to concrete cases. The casebook of actual and projected ethical problems in scientific research is an attempt to provide some experience in analyzing cases and applying ethical

v

principles to them in order that a justified course of action might be proposed.

A bibliography, ranging over the most important articles in the last ten years, is also provided, since it is not within the purview of this book to deal with all questions or to consider all opinions which may be offered. In addition, the bibliography should provide an invaluable research tool for those who are presently working on or contemplating working on, these ethical problems.

This book cannot be all things to all men. It suffers from many limitations, some of them serious. It was begun from "scratch," and it suffers from all of the defects inherent in a beginning of research. However, if my work is improved upon by others, its purpose will be deemed accomplished.

Secondly, although this work was supported by the U.S. Office of Education and Kansas State University, the whole undertaking proved much more difficult than originally thought. Thus, this book was produced under the dual strictures of a lack of time and a lack of money. However, without the support that was given, the enterprise of studying the ethical problems in scientific research would still be merely a proposal and not a reality.

Thirdly, in this book I have tried to strike a balance between a theoretical study of ethical principles and a casebook of practical applications. Those already knowledgable in ethics may think my treatment of ethics too elementary, and those interested primarily in practical applications may consider the casebook too short. I have not tried to write a strictly theoretical treatment of ethics for ethicians or philosophers, nor a comprehensive casebook for casuists, but rather a combination of both. In any compromise, both sides lose a little.

The last limitation will disappoint many, but will alleviate the fears of most. It is *not* the aim of this book to make ethical decisions. This is the sole responsibility of the moral agents involved in any case. What *is* intended is that scientists, who are daily called upon to make ethical decisions in new and complex cases, will know *how* to make them. Thus, the logic and structure of ethics is presented in order that the reader may construct his own ethical theory expressing his own ethical values and justify his own ethical decisions. Those who expected someone else to assume their responsibility will be disappointed, but I am happy to assure the rest that I do not set myself up as the final arbiter of all moral problems arising in scientific research.

Lastly, this work should be seen as the beginning, not the end of ethical research in this field.

If these limitations are kept in mind, we will not ask for what cannot be given. But this does not keep us from demanding all that was promised.

C.E.R

Manhattan, Kansas

ACKNOWLEDGMENTS

It is fitting here to list my debt of gratitude to those who contributed to this project: The U.S. Office of Education, Department of Health, Education and Welfare, whose farsighted policy of supporting research in new fields allowed them to support with money a project which others were willing to support only with words; Kansas State University, and especially Dean William Stamey, who provided me with the time to conduct much of this research, and the encouragement to persevere in it; Mr. John Wagner, Graduate Research Assistant in the Department of Philosophy, for providing the bibliography, and Miss Vicki Voth and Mrs. Kathy Tubbs for typing the manuscripts and for their patience with the erratic schedules with which they were required to work.

I want also to make known my deep appreciation to those scientists who generously gave me almost inordinate amounts of their time: Dr. Val Woodward, Professor of Genetics, University of Minnesota; Dr. Raymond Myers, Chairman, Department of Chemistry, Kent State University; Dr. Thomas Dao, M.D., Research Professor of Physiology and formerly Chairman of the Ethics Committee at Roswell Park Memorial Institute, Buffalo, N.Y.; Dr. J. A. Van den Akker, Professor of Chemistry, Institute of Paper Chemistry, Appleton, Wisconsin; Dr. Lawrence Cranberg, Professor of Physics, University of Virginia; Mr. J. K. Craver, Chief Research Associate, Monsanto Chemical Co., St. Louis, and Dr. George A. Lewis, Jr., Professor of Psychology, Wichita State University. This whole project evolved from many conversations with Mr. Craver and Dr. Lewis, and its completion is, in large part, due to their continued encouragement and assistance.

I should also like to thank Professor Philip Alger, Rensselaer Polytechnic Institute, for his critical, yet sympathetic, reading of an earlier draft.

All of these people, I am sure, will not agree with everything

I say in this book. Any virtues it has I attribute largely to their assistance; the errors, limitations, and mistakes I keep for my own.

The research reported herein was performed pursuant to a grant with the Office of Education, U.S. Department of Health, Education, and Welfare. Contractors undertaking such projects under Government sponsorship are encouraged to express freely their professional judgment in the conduct of the project. Points of view or opinions stated do not, therefore, necessarily represent official Office of Education position or policy.

CONTENTS

PART THREE

Bibliographies

ETHICS FOR SCIENTIFIC RESEARCHERS

Part One

Introduction to Ethics

I

INTRODUCTION

The purpose of this book is twofold; first, it is intended as an introduction to the theoretical study of ethics for those with no previous training in ethics or philosophy, and secondly, it is intended to provide an opportunity for the practical application of normative principles to concrete cases of ethical problems in scientific research.

In an effort to accomplish these goals, the book is divided into three sections. The first is an introduction to ethics. In this section, the nature of ethics is considered first, followed by a brief account of some of the most popular normative theories. In addition, a portion of this section is devoted to the justification of these normative theories in view of the prevalent meta-ethical positions.

In the latter part of this section I will propose a model of an ethical theory which, if not immune to criticism, at least avoids the criticisms which can be levelled at many other theories. Also included are a brief discussion of values and a short treatment of the question of freedom.

In the second section, the casebook, fifty-five cases of scientific research which raise ethical questions are given. The method of case analysis is discussed and an example of a completely worked out case analysis is given. The cases are loosely grouped into three parts. The first part presents cases of ethical issues which must be decided by the individual researcher. The cases in the second part require the joint analysis and solution of the investigator and his employer. In the final part, ethical problems which affect society at large and which must be decided by society at large are presented.

The third section, an annotated bibliography, is, in my

opinion, a very important aid in the achievement of the goals of this book. Since this work is intended as a beginning of research, the bibliography will provide an invaluable aid both to those doing case analysis and those wishing to do further research on ethics and science.

Because of the dual goals of this book, the theoretical considerations in the introduction to ethics will be held to a minimum. This should be recognized at the outset so that those expecting a complete course in theoretical ethics will not be disappointed and those benefiting from the abbreviated account here will not think that they have mastered in depth the intricate and complex topic of normative ethics. The account of ethics offered here is a compromise between the parameters of simplicity and completeness.

In teaching by the casebook method, the emphasis is on the technique of case analysis and practicum, that is, on the reader's ability to perform case analyses himself.

It is quite possible for someone who completely disagrees with my account of ethics to still benefit from these materials by substituting his own introduction to ethics for mine and then utilizing the casebook and the bibliography.

I have made considerable effort throughout this introduction to clearly distinguish between my own personal opinion and accounts of historical positions. Furthermore, in presenting the latter every effort has been made to be accurate and to present as strong a case as possible.

II

ETHICS

E thics is the study of *conduct*. This means that it is concerned only with that portion of human behavior susceptible to the ascriptives "right" and "wrong." Thus, the behavior of animals, infants, and insane persons is not considered to be conduct. Furthermore, not all adult human behavior comes under the purview of ethics.

Ethics is concerned only with interpersonal behavior. There would be no such thing as ethics if each of us were locked in isolation in our own private worlds. But this does not mean that all interpersonal behavior is moral or immoral. There are many actions and situations which are morally neutral. Unfortunately, it is not possible to lay down hard and fast rules for deciding when and under what conditions behavior is susceptible to ethical ascriptives. In general, however, any intentional action which affects the well-being of others is within the scope of ethics. Practically speaking, our vagueness here is not critical since we all know, in general, to what persons, actions, and motives it is meaningful to apply such ethical terms as "right," "good," "ought," "virtuous," and so forth.

Some confusion does arise, however, in distinguishing among *morals, law and custom* (or *etiquette*). This is understandable since in many cases they overlap and often the same terms are used when speaking about a custom and a moral action, for example: "That is the right way to hold your fork," and "That is the right way to treat your employees." Another example might be the following: "It is forbidden to have an abortion." This might mean it is illegal or it might mean it is morally wrong.

In certain respects, law, morality, and custom resemble one another. Law and morals are similar in that they deal with the more important of interpersonal relations and actions, actions which affect our general well-being. Custom is primarily concerned with more minor matters such as table setting, greetings, reception lines, and the opening of car doors for ladies.

With respect to their initiation and promulgation, moral rules and the rules of etiquette are quite similar. No one person or deliberative body lays down moral rules or customs. They just seem to grow and evolve with a society. Laws, however, are always traceable to individuals or groups who are impowered to set down laws, promulgate them, interpret them, apply them, and change them. Thus, we have the 1964 Civil Rights law made by the U.S. Congress but we do not have the "1923 Fork-Holding Rule" or the "Kansas Law Forbidding Deception During Courtship."

In many cases, these three types of rules of conduct overlap. For example, murder is not only morally wrong, it is also illegal. Another example of human action where all three types of rules apply would be a wedding. There are civil laws prescribing how the action must be performed, there are moral considerations (e.g. one ought not marry a girl solely in order to become heir of her father's estate), and there is a jungle of etiquette rules governing a wedding.

It should not be surprising to find such an overlap. All three types of rules are concerned with human conduct in interpersonal situations. But each type of rule governs from a different perspective. Law is concerned with civil order and the protection of rights; ethics is concerned with morally correct action, while custom deals with socially correct action. It is also easy to present cases where law, morality, and custom are at variance with one another. For example, casual divorce is both legal and customary in the United States. The morality of such action is questionable. Again, it is customary to treat blacks as inferior in certain parts of this country, although it is immoral and illegal to do so.

If the important differences between these three types of rules is kept firmly in mind, few confusions ought to arise. It is just such confusion which is responsible for the ethical views of certain cultural relativists who mistake morality for custom.

Prohibition in this country resulted from confusing legality with morality.

It is also very important at this point to distinguish between ethics and other allied studies of human behavior. Many otherwise competent persons have fallen into a morass of confusion, nonsense, and falsity because they did not carefully appreciate the difference between anthropology, sociology, psychology, and ethics. Today, it seems to me, the confusion between psychology and ethics is central. Psychology is the study of how people actually behave in different situations. Ethics, put simply, is the study of how they *ought* to behave. Thus, except in the limited manner which will be explained in detail later, the findings of psychology are not germane to ethics. Failure to see this distinction has led in the past to the absurd position that in all cases, people *ought* to behave exactly as they *do* behave.

It seems clear to me, however, that as unfortunate as it may be, we often behave differently than we think we ought to behave. Thus, it makes perfect sense to describe how we ought to behave in spite of what we know about our actual behavior. In short, ethics is concerned with *ought,* while psychology, anthropology, and sociology are concerned with *is.*

Normative and *meta-ethics* denote different subject matters and different perspectives. Normative ethics is the study of the norms or criteria by which we judge actions right or wrong. Meta-ethics is the study of ethical discourse. Briefly, normative ethics is the attempt to rationally determine what we ought to do in a particular case, while meta-ethics is the study of what we mean when we say "X ought to do A".

We can further distinguish between these two inquiries by looking at the type of question each attempts to answer. Normative ethics answers such questions as, "What ought I to do in situation S?" "Was he right in doing that?" "What are my moral principles?"

Meta-ethics tries to answer some of the following questions: "What do we mean by our ethical terms, such as 'good'?" "Do moral judgments state facts, express emotions, or give commands?" "How can we justify our ultimate moral principles?" "Do ethical arguments employ a special type of logic?" "Must we universalize our moral judgments?"

Certainly, there is a relationship between our answer to one type of question and our answer to the other type. For example, if I answer the question, "Do moral judgments state facts, express emotions, or give commands?" by saying that moral judgments are disguised ventings of emotions or are merely statements of taste, then it does not make sense to try to answer the normative question "Can I justify my judgment . . .?" The latter question simply does not make sense, given the answer to the first. Matters of taste and expressions of emotions are the kinds of things that cannot and need not be justified.

As a consequence of this relationship, an adequate account of normative ethics must attempt to answer certain meta-ethical questions. It simply is not possible to so separate these inquiries that we can completely answer normative questions without answering meta-ethical questions. It is possible, however, to merely answer the meta-ethical questions and suspend all judgment on the normative issues. This is, in fact, the prevalent trend in contemporary Anglo-American philosophy. However, this attitude is not open to us since the whole purpose of our effort is to answer the normative questions. As a consequence, this account of ethics will deal with both types of questions.

The purpose of an ethical theory is to establish a coherent interrelationship among the elements of the theory. The goal here is ultimately to justify—to show reasonable—particular moral judgments. Ethical theory, like any other theory, is an attempt to show coherently, consistently, and, as far as possible, completely the relationships among moral judgments, moral rules, and moral principles so that the moral rules serve to justify the moral judgments and the moral principles serve to justify the moral rules. Thus, in a well-developed theory, we can show our moral judgments to be correct and reasonable in the light of the theory's rules and principles. The following two sections will deal with this question in greater detail.

Elements of a Normative Theory

The basic elements in any normative theory are four.

(1) The particular moral (normative) judgment. This is simply a judgment that a certain action is right or wrong for a specified individual in a given situation. In addition, we also make normative judgments about persons, e.g. "He was a virtuous man," about motives, "He did that from the admirable motive of benevolence," and about intentions, "He intended to do the right thing, but everything went wrong." It is important to clearly understand that the normative judgment is singular in the sense that it refers to a single action, done or contemplated by a particular individual, in a specified situation.

(2) Moral rules generally refer to classes of individuals, classes of actions, and classes of situations. An example here would be, "Students ought not to cheat on examinations." Compare this with the particular judgment, "It was wrong for John Doe to cheat on his logic exam." Rules not only specify classes of agents, actions, and circumstances, they also often include the exceptions to the rule. Here an example might be, "A person ought not to lie to another person, except when the other person clearly does not have a right to know the answer to his question or when very grave consequences will result from telling the truth, such as endangering national security or causing another person irreparable harm."

The point of particular judgments is to apply the rules to specific cases, while the point of the rules is to apply the principles to specific *classes* of cases.

(3) Moral principles are the most general of moral judgments and represent the ultimate moral commitment of those holding the principle. An example of an ultimate principle might be, "One ought always to do that action open to him which will produce the maximum net expectable welfare for the most people." From this, one could argue that students ought not cheat on examinations because of the deleterious effects cheating has on other students and because in the long run the cheater will suffer from his action. From the rule, we can further particularize our judgment to "John Doe ought not cheat on his logic exam."

(4) The fourth element in ethical theories are second order principles which specify which principle takes precedence when there is a conflict between principles and which rules take precedence when there is a conflict between rules.

Consider the following example of possible conflict between principles. P_1 One ought always maximize net expectable welfare. P_2 Welfare should be distributed justly, that is, according to relative need. Now suppose there are two possible actions, one which maximized the net sum of welfare (when "welfare" is uninterpreted, it could be pleasure, material wealth, contentment, intelligence, etc.) and one which produced less welfare, but what was produced would be more justly distributed. What action, then, ought someone subscribing to both of these principles do? The purpose of a second order principle is to adjudicate just such cases. As an example, a second order principle here could be, "When P_1 conflicts with P_2, P_1 takes precedence."

More frequent are cases where rules conflict. For example, suppose I promise to repay a loan on a certain date and on that date I have the money to repay it. But the lender comes to me and says he wants the money to hire someone to kill his wife. Here we might have the rules, "One ought to keep his promises," and "One ought to prevent harm from coming to another person if it is within his power to do so." In this case, most of us would, I suppose, refuse to honor our promise to repay the loan. This shows that we generally subscribe to a second-order principle which states that R_2, preventing harm, takes precedence over R_1, keeping promises.

Of course, a normative theory with only one principle does not have to face the problem of principle conflict. But whether or not it is possible to have an adequate theory with only one principle is a question which will be discussed later. In general, then, normative theories must have second-order principles (and the ranking of principles is in effect a second-order principle) to adjudicate cases of conflicting principles.

The Criteria for Assaying

There are four criteria for assaying a normative theory: consistency, universalization, relative completeness, and parsimony. Before elaborating on each of these criteria, it is indispensable to discuss the notion of assaying.

Normative ethics received a "bad press" during the first half of this century because of confusion between "verifying," "confirming," and "assaying." It was discovered—and the discovery should not have come as any surprise—that it is not possible to verify (demonstrate as true) ethical principles and ethical theories. For, in the strictest sense, only tautologies can be verified and only contradictions can be falsified. A tautology is verified since it is logically necessary that it be true.

Ethical principles, however, cannot be tautologies since the latter merely exhibit the logical relationships between concepts and cannot prescribe a course of action. Nothing follows about how one should conduct himself from a tautology. Thus, if ethical statements were tautologies, they could not function as ethical statements. If they are not tautologies, they cannot be verified.

A weaker sense of "verify" is synonomous with "confirm". The statements of empirical science, since they too are not tautologies, cannot be verified. But they can be confirmed or disconfirmed. Confirmation is an estimate of the probable truth of a statement based on a given set of evidence. For example, based on the evidence we now possess, the statement that "cigarette smoking is causally linked with lung cancer" is confirmed. This does not mean that relative to another set of evidence the statement will not be disconfirmed. Thus, since confirmation is always relative to evidence, it is a first principle of scientific methodology that any adequate confirmation or disconfirmation must be relative to *all* of the available evidence. If an investigator picks and chooses his evidence carefully—and this is sometimes done—almost any empirical statement can be confirmed relative to the evidence.

But the notion of confirmation does not easily apply to ethical statements and theories because of the difficulty in determining what would count as evidence. Since ethical statements are "ought" rather than "is" statements, does any given set of actual empirical evidence weigh in favor of or against them? For example, what psychological evidence could be called to confirm or disconfirm the statement "One ought always do that action open to him which will produce the maximum net expectable welfare"? Ethics is not concerned with what people *think* they ought to do, but with what they *really* ought to do. Thus, an opinion questionnaire is not germane to establishing or discrediting an ethical statement. For we all recognize that people have often thought that they ought to do things which are in fact immoral, e.g. the Germans thought they ought to purify their race.

As a consequence of recognizing that ethical statements cannot be verified (since they are not tautologies) and they cannot be confirmed (since the notion of what would count as evidence is so nebulous here), many philosophers concluded that there was no way of evaluating ethical discourse. This conclusion is, I believe, quite false. The correct conclusion is that ethical statements cannot be evaluated as logical and empirical statements are evaluated. This does not mean that they absolutely *cannot* be evaluated.

In my opinion, an ethical statement is evaluated only in relation to the ethical theory in which it plays a part. In isolation, a single normative statement cannot be evaluated. But this does not differ substantially from the case of empirical statements in this respect. No empirical statement is confirmed or disconfirmed in isolation. Its confirmation is always relative to the theory in which it plays a part. Therefore, in the final analysis, it is the theory which is directly evaluated and the statements in it that are indirectly evaluated.

It may seem that I am working this point too hard, but this is not the case. Failure to understand the notions of confirmation and verification led the logical positivists to both an erroneous philosophy of science and to an erroneous meta-ethics. The criteria for evaluating ethical theories is just slightly different from those used in evaluating other kinds of theories. For this reason, I use the term "assay" to indicate an evaluation of ethical theory which differs from the procedures of verification and confirmation.

In assaying normative theories, we use first the criterion of *consistency*. In this respect, ethical theories do not differ from other theories. For, an inconsistent theory (one having contradictory principles or principles from which a contradiction can be inferred) is simply not a theory. The whole goal of a theory is to rationally account for some set of experiences. An inconsistent theory is simply not a rational account. It is not rational since it violates a cardinal logical rule of system building and it is not an adequate account of any set of experiences since contradictory statements about those experiences can be inferred.

On the theoretical level, this criterion seems quite correct and indisputable. So much so, perhaps, that one wonders how anyone would ever be led to violate it. But in actual practice, inconsistent sets of normative principles are quite common. For example, suppose someone had as his normative principles the following: P_1 All men are brothers and should treat one another accordingly. P_2 Negroes are inferior and should be treated accordingly. In addition, this person might agree to the factual statement that "Negroes are men."

This set of normative principles would lead to the following contradiction: Negroes are men; all men are brothers (i.e. equals); therefore, Negroes are brothers and should be treated accordingly. Now with this conclusion and P_2 we get the contradiction that Negroes are brothers and Negroes are not brothers (i.e. are inferior). The only logically possible way to avoid the contradiction is to argue that all men are inferior. But this is an absurdity since if we are all inferior, then the term "inferior" becomes meaningless.

This may appear to be a rather simplistic example, but I think that it is more common than we perhaps realize. In general, people do not recognize that they hold an inconsistent set of normative principles because they never attempt to formalize their principles and look at the whole set. In everyday life, we usually call upon only one principle at a time and may fail to see that the principle we relied on yesterday contradicts the principle we are relying on today.

One of the purposes of normative ethics is precisely to point out such inconsistencies. When a person does hold contradictory normative principles, he must give up one of them or refuse to be rational. If the latter, then he has taken himself and his normative beliefs out of the realm of ethics. Thus in ethics, normative principles must be consistent with one another and an inconsistent ethical theory is rejected for this reason alone.

The second criterion by which we assay ethical theories is *universalization.* Universalization is simply the principle that what is right for me to do is right for anyone like me to do in like circumstances. Thus, if I want to argue that action A is right for me but wrong for you, I must exhibit a relevant difference between you and me, or between your circumstances and mine, or both.

Here the problem has often arisen in determining what are relevant differences. Certainly, you and I are different persons, we have different histories, different names, different families, and occupy different spatial locations. Thus, it has been argued, universalization is absurd because I must necessarily be different from you and my circumstances must necessarily be different from yours. It is for this reason that we insist on *relevant* differences.

The case here is very much like that in science or in any generalization. This pan of water is different from that pan of water: it is in a different pan, occupies a different spatial location, etc. But does this invalidate any generalizations about the properties of water? No, for we all understand that these differences necessarily exist. There are, however, relevant differences which make a statement about one pan of water not apply to another pan of water. For example, this pan of water is salt water while that one is fresh. This pan is a pressure cooker while that is a sauce pan. Now these differences are relevant because a generalization about the temperature at which pans of water boil will apply to one pan while not to the other. We might say here that these are chemically relevant differences since they affect the chemical properties of the respective pans of water and their boiling temperatures.

The same holds true in ethics. We all understand that there are some necessary differences among agents, actions, and circumstances, but these are not ethically relevant. These differences do not invalidate our normative generalizations and do not render ethics a list of singular judgments. If we can carry our analogy with chemistry a bit further, we can say that ethically relevant differences are those which affect the moral properties of the agent, action, or circumstances. In other words, ethically relevant differences are those which would lead us to change or modify our normative judgments.

Since in our normative prescriptions and generalizations we make an effort to specify the exact nature of the circumstances and of the action (i.e. we give the ethically relevant aspects) and since we presume that all men are basically alike with respect to certain moral properties, we insist that our normative generalizations apply to all men doing the specified action in the specified circumstances unless we can be shown some ethically relevant difference between most men and the agent who wishes to exempt himself from the generalization. This is what we mean by saying that normative judgments must be universalizable.

If ethical judgments were not universalizable, they would be of no more use than scientific statements about the behavior of water if the latter were not universalizable. Furthermore, it is the whole purpose of theories to talk about classes of things, and although some classes have a single member, if all classes had only one member, the whole idea of theories, principles, and generalizations would be absurd.

This criterion, like the first, appears indisputable, I suppose, until we see a practical application of it. Ethical egoism is rejected solely on the grounds that it violates this principle of universalizability. Suppose I had the single moral principle: "In all cases, do that action open to you which will most benefit C.E. Reagan." This principle violates our criterion because it can in principle apply only to me. This is not a matter of there just happening to be only one member of the class, C.E. Reagan. There can only be one member. This would differ from the case of a principle such as "In all cases, the President of the United States ought to . . . ." For here it is not logically necessary that this class have only one member, though in fact it does. Secondly, it is not limited to one and only one specific individual—in fact we have had at various times about forty individuals who have been members of that class. The case differs in another respect in that there is nothing specifically ethically relevant about being a C. E. Reagan, while there might be ethically relevant differences between most men and presidents.

The first attempt to avoid this objection usually results in changing the principle to "An agent ought always to do that action which most benefits him." While the principle itself is theoretically universalizable, any action which might fall under the principle is not. For example, from the egoist principle, I might try to justify liquidating you. But I cannot universalize that moral prescriptive since to do so would result in something like this: It is right for everyone like me to liquidate everyone like you. But if you were also an egoist, the proposed action would be, by your egoist principle, an absolute wrong—since liquidation does not usually benefit the person being liquidated. Thus, we would have the same action, my killing you, being both right and wrong and we would appeal to the same principle to argue that it is right (me) and that it is wrong (you). But if the action is right for me and wrong for you, then by the principle of universalization, there must be some ethically relevant difference between you and me. Such a difference is notoriously difficult to exhibit and is not contained in the egoist principle, which begins, "An agent . . . ."

But the egoist's difficulties with universalization do not stop there. An important part of ethics is advising others. Now suppose Jones comes to me and tells me he wants to embezzle from Smith and is sure he will not get caught. Since we are both egoists, do I then advise Jones to go ahead since the additional wealth will be to Jones benefit? Perhaps, but suppose I am part owner of Smith's company (unknown to Jones). Now do I advise Jones to go ahead, since it is to his own benefit, or do I advise him not to, since Jones' embezzling will hurt me? The whole matter of advising another concerning ethical matters becomes a mare's nest of confusion since it is not possible to universalize normative judgments from the egoist's viewpoint.

Exactly the same type of confusion arises when the egoist tries to make normative judgments about the actions of others. Does he judge the action on the basis of how much it helped or harmed him personally or on the basis of how it benefited the agent?

Ethical egoism, then, is rejected because it cannot be universalized, and it cannot be universalized because it postulates an ethically relevant difference between the agent and all other men; a difference which cannot be exhibited.

The third criterion used in assaying normative theories is *relative completeness*. Thus, a normative theory which could not handle sexual morality would be incomplete—and, I might add, in a rather significant way. The whole point of an ethical theory is to account for, in the sense of giving direction to, our moral behavior. But if there is a segment of our moral behavior that is neglected by a moral theory, then that theory is incomplete in that respect. Now, completeness is, in my judgment, relative since it may be impossible for a theory to be absolutely complete and to, at the same time, fulfill all of the other criteria. Completeness is relative in another sense also; that is, with respect to other theories. If one of two competing theories is more complete then, on the basis of this criterion, it is preferable to the less complete theory. Thus, theories are said to be incomplete only with respect to other theories.

There are two points to add here. First, some theories have attempted to be complete by having only one very general principle. A notable example is agapism whose only principle is "In all cases, do the loving thing." It is complete in the sense that it covers all possible situations, but it is incomplete in that it does not give the criteria for "the loving thing." I think that if specific criteria for "the loving thing" were given, we would find that its completeness is only apparent. Another example of attempted single principle completeness is the utilitarian principle, "Always do what will produce the maximum net expectable welfare." For, even if we interpret "welfare" we will find that the principle is incomplete for in some cases it cannot adjudicate between very different proposed actions. In our later discussion of utilitarianism I will argue that the principle of justice or distribution must be added to the principle of benevolence.

This is not a theoretical point, but it is interesting to note that for many years morality in this country, while not theoretically so incomplete, in practice was grossly incomplete. I am specifically referring to the tendency to equate morality with sexual morality and to ignore other important areas of ethical behavior (e.g. business practices). This sort of incompleteness was not the fault of the theory, but of those who applied the theory.

The last, and in my opinion, the least important criterion for assaying ethical theories is *parsimony*. Parsimony or simplicity as it is sometimes called, means the theory having the least number of principles is to be favored. All other things being equal, the theory with one or two principles would be favored over the theory with seven or eight.

In practice, this criterion is not too important for several reasons. In the first place, it is always logically possible to reduce any number of principles to one principle by connecting the principles with conjunctions. Secondly, most theories with one or two principles tend to be so vague that the principle is inapplicable or the theory is incomplete. The example of agapism comes to mind here. Thirdly, it is very difficult to justify this criterion except on aesthetic grounds. One consideration that does seem to enter in here is that the fewer principles, the less likely there is to be inconsistency. For these reasons, parsimony is the least important consideration in assaying normative theories and should be applied only after all of the other criteria have been applied.

So far we have been discussing only the general nature of ethics and normative theories. The purposes and the elements of normative theory are the same for all theories. The criteria I have outlined are used in assaying all theories. In the next section, we will discuss some of the more popular normative theories which have been offered, many of which are widely held today.

III

SOME NORMATIVE THEORIES

In this section, I will give a brief account of several normative ethical theories which have been popular at one time or another. This treatment of normative theory is not intended to be a detailed historical account nor does it pretend to be complete in the sense of covering all of the possible normative theories. Rather, I have chosen the theories presented here in order to illustrate some of the important differences among types of normative theory.

Theories are often classified according to two criteria: *Teleological* (*deontological*) and *rule* (*act*). The first criterion distinguishes between theories which maintain that the most important feature of a normative principle or action is its result, and theories which maintain that the most important feature by which we evaluate principles is the nature of the act or class of acts involved. If, for example, we judge actions on the basis of what type of action it is, then we are evaluating them from the deontological perspective. If, on the other hand, we are evaluating actions on the basis of the state of affairs the action produces, then we are taking a teleological point of view.

This distinction can be clarified by looking at the following example: Suppose I have made a solemn promise to someone to do action A. Later, I realize that by not doing A, the consequences will be much better. The deontologist would advise me to keep my promise and do A, since keeping one's promises is the kind of an action that ought to be done. He would justify this advice by pointing to the principle that we ought to keep our promises. The teleologist, on the other hand, would advise us to do the action which would produce the best effect, and if this meant breaking a promise, then we should break the promise. (Of course,

21

our teleologist is including the effects of breaking the promise with the total effects of not doing A.) In sum, then, the teleologist is evaluating actions and principles on the basis of their effects, while the deontologist is evaluating them as to what kind of an action A is. The deontologist may take the effects of an action into consideration, but this is not the sole or the most important consideration. For the teleologist, it is.

The second classificatory distinction is whether we judge an action on the basis of that particular action in those particular circumstances or judge the action on the grounds that it is a member of a particular class of actions. Thus, an act-teleologist would judge action A on the basis of the effects of that particular action. A rule-teleologist evaluates A on the basis of the effects of having a rule obliging us to do the whole class of actions A. Act-deontologists evaluate actions according to the type of action A is, while rule-deontologists look to the nature of the whole class of actions of which A is a member. These distinctions will, I think, become clearer as we see theories which exemplify the various combinations of these criteria.

Situation Ethics

Situation ethics is an act-deontological (though it can be teleological) theory whose main tenet is that we cannot have general rules as to what everyone should do in a certain situation.[1] According to the situationists, every action and every set of circumstances is unique. For this reason, no general rule can be universally applied. They lament the "rule idolatry" of most normative theories and insist that only by a careful examination of all of the relevant factors in any given case can we decide what is the right thing to do.

This theory stems largely from the writings of the existentialist philosophers who object to rule theories on two grounds. First, each person is unique. There is no "human nature" such that what is right for one man is right for all men. Each man must strive to be himself, to be authentic, and not allow himself to be stuffed into a mold. Furthermore, each man chooses his own values. They do not come down from heaven, nor can they be established once and for all. Rather, we choose our values in relation to the person we are and to the ineluctable situation in which we find ourselves.

Secondly, since we are unique, any situation in which we find ourselves is unique. We cannot deal with men or with circumstances by using class logic. For to put individuals in classes is to treat them as though they are not unique.

Much contemporary Christian theology is imbued with existentialism and situation ethics has become popular with the younger clergy. They object to the traditional church's absolute rule orientation and insist that the rules simply do not apply in all cases. For example, traditionally the Christian church has objected to any form of premarital sex. Yet the situationists argue that this prohibition is simply too broad and too absolute and that much harm comes from it. Their alternative is to say that each particular case of premarital sex must be examined as a unique case and that a judgment must be rendered on the basis of the factors involved in that particular case. Sometimes this will require that the rule be violated; but, the situationist says, more good results in these cases from not adhering to the rule than would result if the rule were ploddingly followed.

Situation ethics, then, is considered an act theory since each particular action is evaluated on its own merits and without reference to any rules of obligation. It may be a teleological theory if the main consideration in evaluating the particular action is the effects or consequences of that action. However, most proponents of this theory judge each action on the basis of the kind of action it is. ("Irrespective of the consequences, it is just not right to do that.") Thus, I have classified situation ethics as an act-deontological theory.

Those who seek rules to guide their actions are usually frustrated by situation ethics since its main principle is something like "Just examine the case and do what you think is right."

A typical objection to situation ethics is the following: "Particular moral judgments are not purely particular, but are implicitly general." This means that if I decide that in this particular situation doing A would be the right thing, I am implicitly saying that anyone in these circumstances ought to do A. If we look at a more concrete example, this seems to be the case:

Suppose that I go to Jones for advice about what to do in situation Y, and he tells me that I morally ought to do Z. Suppose also that I recall that the day before he had maintained that W was the right thing for Smith to do in a situation of the same kind. I shall then certainly point this out to Jones and ask him if he is not being inconsistent. Now suppose that Jones does not do anything to show that the two cases are different, but simply says, 'No, there is no connection between the two cases. Sure they are alike, but one was yesterday and involved Smith. Now it's today and you are involved.'[2]

Such a response from Jones would strike us as absurd for the reasons that I gave when discussing the principle of universalization. We all agree that there are some necessary differences between the two cases, but that is not important. What is important is whether or not there are any ethically relevant differences between yesterday's case and today's case such that different ethical judgments are called for. Here, the principle of universalization is being used as a reason to object to situation ethics.

The situationist, however, may quickly come back with the argument that the differences between Smith and me are ethically relevant because we are totally unique individuals and, as a consequence, no moral prescription that applies to him need necessarily apply to me. The situationist's grounds here for saying that we are unique are no longer the simplistic reasons that we occupy different spatial locations and we have different names. He is advancing a metaphysical argument concerning the nature of human persons. Men do not have a unique human nature, rather it is human that each man is unique.

It is not my purpose here to debate the metaphysics involved. It will be enough to point out the consequences of this position. There can be no such a thing as ethics, and, in fact, this is the conclusion of most existentialists.[3]

For myself, however, I do not accept this conclusion for two reasons. First, there would be no justification for continuing with our efforts to understand ethics; all ethical theories would be bogus. Secondly, I can admit that each person is unique in a metaphysically significant way without accepting the inference that it is therefore impossible to universalize moral judgments. People are psychologically unique, but this does not in itself invalidate all of the generalizations of psychology. It seems to me that this uniqueness is understood when we make generalizations, whether they be in psychology or in ethics. Every theory is an attempt to generalize, and every generalization must, by its very nature, overlook the peculiarities of the individuals about which it is a generalization. This should serve as a caution in the use and application of generalizations; it does not by itself invalidate all of them.

Phrased differently, I think that human beings are sufficiently similar to justify ethical generalizations, but in our application of those generalizations to specific cases, we must not forget completely the uniqueness of the individuals involved. Thus, I think we are justified in continuing our theorizing but must be cautious when applying our theories to particular persons, actions, and situations.

Formalism

With the previous caveat in mind and the view that moral judgments are implicitly general, we can now take a look at a rule-deontological theory. Formalism is the view that the standard by which we evaluate the moral qualities of actions is a set of nonteleological rules. These are rules which set down which classes of actions are obligatory and which classes of action are prohibited. In itself, the Decalogue is such a set of rules.

Although not currently in favor, formalism was generally accepted for a considerable period and had in this century a very formidable advocate, W. D. Ross. The essence of Ross' theory is a set of six rules of prima facie obligation:

(1) Some duties rest on previous acts of my own:
    (a) There is an obligation of fidelity
        (e.g. promise keeping, truth telling, etc.)
    (b) There is an obligation of reparation for wrongful injury or harm I have done to others.
(2) Some duties rest on previous acts of others: There is an obli-

gation of gratitude (e.g. returning a favor someone has done us).

(3) There is an obligation to see that happiness or pleasure are distributed according to merit.

(4) There is an obligation to improve the condition of others with respect to virtue, intelligence, and pleasure.

(5) There is an obligation of self-improvement with respect to intelligence and virtue.

(6) There is an obligation not to harm others and to prevent harm coming to others.[4]

As I mentioned in our discussion of consistency in ethical theory, any time there is more than one principle there is the possibility of conflict. What ought we to do when rule (1) conflicts with rule (6)? One possibility is that we rank our rules in a hierarchy such that in any conflict the rule with the lowest number takes precedence. This would be the clearest and easiest method of adjudicating conflict. But Ross and others have rejected this method because they believed that it was not possible to make such a hierarchy of rules by other than completely arbitrary means. It simply is not clear which is the most important rule, the next most important, and so on.

Instead, Ross introduced the notion of prima facie obligation. This means that there is an obligation of fidelity if there are no other ethically relevant considerations intervening. When rules conflict—and this is a case of intervening ethical considerations—then we must refer to a second order principle. That act is one's duty which is in accord with the more stringent prima facie obligation.[5] In some cases this seems fairly clear. We have an obligation to keep our promises and an obligation not to harm others. If a case arose where both of these obligations could not be fulfilled, we would have to choose the more stringent obligation. Here we are helped by Ross' statement that the act is one's duty which has the greatest balance of prima facie rightness over prima facie wrongness.[6] Suppose in our case of conflict, my keeping my promise resulted in hurting someone's feelings. We would probably feel that our obligation of fidelity was stronger. If, however, my keeping my promise cost someone else his life, then our obligation not to injure others would be more stringent, in which case we would be justified in breaking our promise.

Admittedly, in many cases it may not be this easy to decide which of our prima facie obligations is our actual obligation. But a formalist could plausibly argue that where it is impossible to tell which obligation is more stringent or which act will produce more prima facie good, then we may do either action, follow either rule.

There are two other problems with formalism. First, it is very difficult to tell when our list of rules is complete.[7] Shouldn't there be a rule in the list dealing with premarital sex? One reply might be that this is covered in the rule obliging us to do good for others and the rule obliging us not to injure others. But if completeness is gained in this manner, that is, by stating the rules in very broad terms, then there is the danger that the rules will not be specific enough to serve as practical guides.

The second difficulty has to do with exceptions to the rules. We all accept, I suppose, our obligations to keep our promises and to be truthful. But we all would insist on a few exceptions to this rule. For example, must a prisoner of war be truthful with his interrogating captors? And, may we not sometimes break a promise when keeping the promise would produce dire consequences?

Ross' method of handling exceptions is the same as handling conflicts of rules. Our rules give us only prima facie obligations, not actual obligations. Exceptions are decided on the basis of the two second order principles exactly as we adjudicate conflicts by these principles. For example, our prisoner of war may be justified in lying because the rule obliging us not to harm others is more stringent than our obligation of fidelity, and his telling the truth would result in injury and harm to other soldiers.

Another way of handling exceptions is to include them in the rules. This can be done either by enumerating the permitted exceptions (e.g. we are obliged to keep our promises except when someone else's life or limb would be put in jeopardy, except when . . . .) or by phrasing the rule in such a way that it will admit the exceptions without becoming a sieve. An example here might be, "We are obliged to tell the truth only to those who have a right to know the truth." Obviously, the rude inquisitor has no right to know personal matters about us and the captor has no right to expect the prisoner to tell him where the rest of his men are hidden. However, it is quite difficult to phrase the rules in such a way that they can satisfactorily do this. Secondly, we are stuck here, as we are with conflicting rules, with having to decide who has a right to know.

In summary, then, the strongest point of formalism is that as a theory it most nearly adheres to the way we actually make moral judgments, that is, by reference to nonteleological rules. Its weaknesses are several. How can we tell when the list of rules is complete? What are we to do with conflicts of rules? What are we to do with exceptions? These, I might add, are not minor defects since the most difficult cases in ethics arise precisely when our list of rules is incomplete, there is a rule conflict, or when we are dealing with an exception.

A major objection to all deontological theories is that it does not make sense, according to the teleologists, to follow a rule when it is clear that a better state of affairs would result by breaking the rule. Furthermore, they insist that it is by the consequences of our actions that we evaluate them. Promise keeping is generally good, since it generally produces better effects than infidelity. But it is absurd to keep a promise when it results in a catastrophy. As a result, they insist that it is the consequences of an action which are the prime consideration in evaluating them.

Other than agreeing that only a teleological theory can ultimately be acceptable, teleologists agree on little else. Just as with the deontological theories, teleological theories can be divided into act and rule theories. Thus, let us first examine act-utilitarianism.

Act-Utilitarianism

The fundamental thesis of act-utilitarianism is that actions are to be evaluated by their consequences, that is, by their producing a greater balance of good over evil in the world.[8]

Furthermore, in our evaluation, we look only to this particular action being performed by this particular person in these particular circumstances. We are not concerned with classes of actions or classes of agents, only with particular actions and particular agents. The question for the act-utilitarians is what would be the result of this agent doing this action in these circumstances, not, what would be the result of everyone doing this kind of action in this kind of situation.

The basic normative principle of act-utilitarianism is, "Do that action which, among all of your alternatives, will produce the greatest balance of good over evil." This has also been expressed as, "Do that action open to you which will maximize the net expectable utility." From the point of view of normative theory of obligation, we can leave "good" and "utility" uninterpreted for the time being. Historically, such things as pleasure, self-realization, material well-being, happiness, and eternal salvation have been used as utility values.

There are three basic problems with this theory, and the discussion of these problems should clarify the theory itself. In the first place, act-utilitarianism supposes that it is possible to somehow quantify good and evil, utility and dis-utility. On the surface, this seems plausible because we often speak of one thing being more pleasurable than another, or of one action producing more material well-being. But upon closer examination, the difficulties are enormous. Of all of the interpreted utility values, pleasure would be the easiest to quantify. Yet, is it more pleasurable to drink a bottle of Nuits St. Georges, 1959, or to see a sunset from the top of a high mountain? Is lovemaking more pleasurable than either of these? Is it more pleasurable for a forty-year-old married man than for a college student? Is winning at pinball more pleasurable than correctly solving a difficult problem in calculus? Can we give any of these numerical values and rank them on a hierarchial scale? Thus, I think that with a little reflection, we will come to understand that it is not possible to quantify utility values sufficiently to perform the weighing or calculus required by act-utilitarianism.

The second difficulty is that this theory requires that we

decide the relative balance of good over evil for every specific action in every particular set of circumstances. Theoretically, this is possible. But in practice, we find that we can rarely perform the evaluative task for every contemplated action. Often there is insufficient time before a decision is demanded of us. Often we lack sufficient information as to probable consequences. Thus, the theory asks us to perform a normative calculation for every action, a task that is virtually impossible in practice.

The third difficulty is that in many hypothetical cases—if not in actual cases—we could have two alternatives, A and B. After a careful calculation, if this were possible, suppose we discover that both A and B have the same utility value, that is, either action will produce the maximum net expectable utility or the greatest balance of good over evil. Then, according to the act-utilitarian, both actions would be morally equivalent. But further suppose that action B involves injuring the reputation of someone as well as lying and deceit. These consequences have been taken into consideration, *ex hypothesi*, in the calculation. Most of us, I think, would deny that both actions are right or that they are equally right. Clearly action B is morally wrong.

This objection is extremely strong and has led to a general turn away from act-utilitarianism toward rule-utilitarianism.

Rule-Utilitarianism

Like act-utilitarians, rule-utilitarians insist that we evaluate actions only by their consequences.[9] But they add that we do not evaluate particular actions, but classes of actions. They would not say that this particular case of lying is wrong because of its consequences (because in some cases, the consequences of lying produce more good than evil) but because the class of actions, i.e. lying in general produces less utility than the class of actions, i.e. truth-telling.

Another way of saying the same thing is that the rule-utilitarians assess particular actions by whether or not they accord with a normative rule. The rule is evaluated by the consequences of having the rule. Let us take a famous (perhaps I should say notorious) example, the Jockey Club case: Suppose you are shipwrecked on an island with only one other man, and this man happens to be near death and very wealthy. He offers you a key to a locker in Grand Central Station and you promise him that as soon as you are rescued you will take the $10,000 in cash from the locker and give it to the old man's favorite charity, the Jockey Club. The man dies and you are rescued. What ought you to do with the money?

The act-utilitarian would urge you to consider all of the possible alternatives, such as giving the money to the United Fund, or the Foundling Hospital, as well as the alternative of fulfilling your promise and turning the money over to the Jockey Club. Now as a result of your calculus (and it may not be too difficult in this case) you decide that the maximum net expectable utility will result from your giving the money to the United Fund. In this case the act-utilitarian would say you are quite justified in breaking your solemn promise and giving the money to the U.F. rather than to the somewhat decadent Jockey Club.

The rule-utilitarian would object strongly, insisting that perhaps in this particular case more good would be done by your proposed action. But, he might continue, you would be breaking a promise and it is wrong to break promises. He would justify his advice on the grounds that having a rule of keeping promises produces much greater utility than not having such a rule. With this statement we would all agree, considering the dire effects of general infidelity.

The rule-utilitarian has certain considerations which weigh heavily in his favor. In fact, we do generally follow rules in actual practice. In any ethical situation we look first to the relevant rule or rules. Secondly, contrary to the view of the act-utilitarian, I think that it is important to ask ourselves, "What would happen if everyone did this?" The act-utilitarian would object to our argument saying, "Yes, but everyone will not do this. And, the knowledge that everyone will not violate the rule of promise keeping is part of the relevant information that contributed to my decision."

Although the Jockey Club case is a little far out, there is a closely analogous situation in actual life. For example, take the case of a person cheating an insurance company or bolting on a credit account. He often reasons as does the act-utilitarian, saying that much more good will come of his having money than his not having money (or unpaid-for goods). The insurance company or the department store is so large that they will not be hurt. Furthermore, he is confident that everyone will not cheat these large companies, so no dire effects will result for the whole institution of insurance or of credit buying. Thus, in this particular case he ought to cheat the insurance company or bolt on his bill because his action will produce the maximum net expectable utility.

Here the rule-utilitarian must take exception. The question is, claims the rule-utilitarian, not whether everyone *will* bolt on his credit account, but whether everyone would have the *right* to do so. It would seem that it follows from the principle of universalization that if our man in the above case has the right to bolt on his account or cheat his insurance company then everyone like him has the right. Now he may say, "Yes, everyone who made the same calculation that I did would have the right to do what I did." But this is not plausible. For, if a person has a right to do something, then his doing that thing cannot be wrong. But if everyone has the right to cheat their insurance companies, then they cannot be said to be doing evil if they do what they have a right to do. Obviously, if everyone does, in this case, what he has a right to do, the consequences would be quite severe, destroying the practicality of insurance and credit buying; and producing more evil than good. Any utilitarian would agree to that.

As a result of the above argument, I think that act-utilitarianism is ultimately indefensible. Their position is plausible only as long as we fail to distinguish between what anyone is likely to do and what everyone has a right to do.

Although it is clear that rule-utilitarianism is far superior as a theory to act-utilitarianism, both theories suffer from an additional and very telling objection. Simply stated it is this: the principle of utility does not tell us how to distribute the good that we produce. For, suppose that we have two alternatives,

A and B. Each produces an equal sum of net expectable utility. But in alternative A, all of the utility goes to a small group of persons. In B, the utility is more broadly distributed. The principle of utility cannot tell us how the distribution ought to be effected and thus cannot tell us whether A or B is the preferable course of action.

The utilitarian might counter by saying that either action is morally acceptable, as long as either action produces more utility than any other alternative. But if we look at a practical case, we see that this is just not so. Right now in the United States we are debating the question of the distribution of wealth. Let us assume that no matter how we choose to distribute our wealth, the sum of wealth remains the same. The critical question is whether we will raise the income tax on the wealthy and grant a negative income tax to the poor or whether we will continue as we are now, taxing the poor and exempting the rich. The point is not to debate this particular political issue, but to illustrate the weakness of utilitarianism; it cannot provide a guide for the distribution of utility.

Some utilitarians have argued that distribution (or justice) is included in the principle of utility. But the very fact that we can present a case of conflict between producing the maximum utility and the distribution of a lesser amount of utility, or the case of different possible distributions of the same amount of utility, seems clearly to show that distribution is not included in utility and that we need a principle of justice as well as a principle of utility.

This may not appear to be a significant concession to make on the part of the utilitarian, but it is. For if we add a principle of justice, his theory is no longer purely teleological. As I have already shown, the principle of justice is not included in the principle of utility and thus cannot be derived from it. It is independent of utility. Utilitarians, however, are very uneasy about accepting a deontological principle since, in general, they maintain that the consequences or effects of an action are the most important criteria for evaluation.

Another reason that utilitarians have objected to the addition of the deontological principle of justice to their theory is that they wanted a monoprincipled theory. With only one principle, they can avoid the problems of consistency and conflict. It is my opinion, however, that utilitarianism is a clear case of a theory which is inadequate because of incompleteness. And, it is incomplete in a very significant way.

I would like to advance here an argument that to my knowledge has not been raised before. It seems to me that a utilitarian should not object to having a deontological principle added to his theory since his principle of utility is itself a deontological principle. This may seem puzzling at first since according to the principle of utility, actions are evaluated by their consequences. In this sense, the principle is teleological. The principle itself, however, cannot be justified by the consequences of having such a principle (i.e. it produces more good to have the principle of utility than it would without it) since that would be obviously circular. I think the principle is justified by saying that producing the greatest balance of good over evil is the kind of action that is right, just as not injuring others is the kind of action that is right. I might also note here that the deontological theory of Ross included some principles that are teleological in the first sense, for example, there is an obligation to do whatever good we can for others.

There is one last difficulty that I find with utilitarianism. It does not take into consideration the avoidance of injury or harm to others. Suppose again that we have two alternatives, A and B. By hypothesis, both produce equal amounts of good over evil. But alternative B includes seriously harming an individual while A does not. Can we continue to say that they are equally right? Now rule-utilitarianism can easily handle this case, as I have already shown, by having a rule against harming others. Act-utilitarianism is defective in this case. But what has been suggested in the light of such an example is that we have a negative-utilitarianism which would have as its basic principle not produce the maximum good, but avoid harm and injury to others. The difference is this, on the positive principle, it would be possible to injure others as long as more good than evil were produced.

On the negative principle, injury to others must be avoided, even if this does not produce the maximum good.

Although I am sympathetic with the motives behind negative utilitarianism, I think that it is unnecessary, unless one insisted on being a negative-act-utilitarian. Rule-utilitarianism can handle cases of injury, as can most deontological theories. The answer to the negative utilitarian then, is to give up act-utilitarianism. Finally, if deontological theories can handle injury cases, and I am right in my argument that even utilitarianism is ultimately a deontological theory, there is no reason to have a negative utilitarianism.

A rather extensive consideration has been given to utilitarianism because it is historically important and it is still widely accepted. Thus, let me conclude this account by summarizing the difficulties that I find with the theory. The most serious problem is that it is not possible to quantify utility, good, or value, and such a quantification is required by both forms of utilitarianism. Secondly, utilitarianism is, in my view, ultimately a deontological theory, in spite of protestations to the contrary by utilitarians.

In the following sections, I will present two normative theories which are widely accepted by common men, but which have rarely been accepted by moral philosophers. I am referring to Agapism and Divine Will theories.

Agapism

Agapism is the ethical theory that advances as its basic tenet that there is only one ultimate normative principle and that principle is to love. This view is explicit in the New Testament when Jesus, being questioned by a doctor of the Law concerning which is the great commandment, answers: "Thou shalt love the Lord thy God with thy whole heart, and with thy whole soul, and with thy whole mind. This is the greatest and the first commandment. And the second is like it, Thou shalt love thy neighbor as thyself." (*Matthew* 22:37-39 and *Mark* 12:29-31) In Luke's account, both of these commandments are conjoined into one single commandment. Thus, whether we wish to consider these as two commandments or as one, they are basically the same. They command us to love God and our neighbor.

This does not mean that there cannot be any additional normative principles or moral rules but that the commandment of love is the ultimate principle from which all of the others follow and which serves as the justification for all of our moral rules. In essence, the agapist is saying that we will be fulfilling the Great Commandment whenever we act in accordance with the other moral principles and rules and we will violate the commandment of love whenever we violate one of the lower order prescriptions.

One immediate objection which historically arose when this theory was put forth is "who is my neighbor?" Jesus answered this question with the parable of the Good Samaritan, which more literally means that all men are our neighbors.

Agapism has the advantage of simplicity and as a mono-principle theory, the advantage of consistency. However, in practice, conflicts have arisen between loving God and loving our neighbor. One notable example was the Inquisition, where many people sacrificed their neighbors for what they conceived to be their loving duty to God's orthodoxy. Theoretically, however, there should be no conflict between loving God and loving our neighbor.

The theory does suffer from incompleteness, and in a rather serious way. It is almost impossible in most critical moral situations to decide what is "the loving thing to do." This can be seen by looking at the following, and not totally infrequent situation. An unmarried girl becomes pregnant. What is the loving thing for her and her boyfriend to do? It is certainly not clear that agapism would dictate one course of action rather than another. The principle may eliminate one or two alternatives, such as killing the girl, or having the boy enlist in the French Foreign Legion, but beyond that, it is of little practical help.

A second difficulty arises from the admonition that we should love our neighbors as ourselves. Suppose that one possible action would benefit us but not our neighbor, while an alternative would benefit our neighbor while doing nothing for us. What should we do if we are agapists? Or, a further complication, suppose an action will greatly benefit us, while harming our neighbor, while an alternative will greatly benefit him, though harm us. Are we required to love our neighbor more than we love ourselves?

Agapism, in my view, is a type of deontological theory since its main principle says that "the loving thing" is just the kind of action that is right. Consequences of our actions may be taken into consideration in determining what is, on the whole, the loving thing, but this is not the main or the primary consideration. The main question is whether our contemplated action is a loving action or is not a loving action.

Since agapism is a deontological theory (and even if it is considered *sui generis*) there could be act-agapists and rule-agapists. The distinction would be the same as between any rule theory and any act theory. Contemporary agapism, as espoused by the hippies, is an act-agapism and stems largely from a revolt against rules and legalistic ethical systems.

In my opinion, agapism is an inadequate normative theory because of its incompleteness. It is simply too vague to serve as a practical guide to action. I tend to think of it not as a normative theory at all, but rather as supplying a much-needed motivation or reason for being moral in the first place. Rather than providing us with a guide for what to do in specific situations, it gives us a reason to be moral, to be concerned about the well-being of others, to look to an adequate normative theory for assistance in determining what is morally right and what is morally wrong.

Divine Will

One final normative theory that merits consideration because of its widespread acceptance is called the Divine Will theory. The ultimate norm on this view is what is right is what God tells us is right and what is wrong is what God tells us is wrong. Or, put slightly differently, what is right is what God commands us to do and what is wrong is what God forbids us to do. Judeo-Christian ethics are largely Divine Will ethics. Adam's eating of the apple in the Garden of Eden was wrong not because eating apples is the sort of action that is wrong, but because God specifically forbade Adam to eat the apple. Similarly with some of the commandments. Working seven days a week is not the sort of thing that is morally wrong (deontological) nor does it produce evil consequences (teleological); it is wrong because God commanded us to rest on the seventh day of each week. In the New Testament we see similar indications of divine will. Divorce is not necessarily wrong in itself, nor does it always produce harmful consequences. But Jesus (who speaks with the authority of God) forbade divorce and remarriage under all (or almost all, depending on interpretation) circumstances. Notice that divorce was permitted by God for the Jews under Moses. This permission was rescinded by Jesus. Thus, divorce is right or wrong depending upon whether or not God allows it.

Certainly, God forbade many actions which would be considered wrong on other theories and for other reasons, such as, adultery, murder, thievery, false witness, and drunkenness. He also commanded many actions which would be obligatory on all other theories, such as, honoring one's parents, caring for the needy, showing compassion for the unfortunate, and telling the truth. What is important is not so much the specific command or prohibition, but the justification of the command or prohibition. The principle which is appealed to by Divine Will theorists is the will of God, not the nature of the action or the probable consequences of the action.

The Divine Will theory, as a theory of general obligation, suffers from what I consider to be fatal defects. First, what can atheists, agnostics, and heathens use as a normative guide on this theory? Are we simply to say that they are outside the purview of ethics, that they simply do not have any way of determining which actions are right and which actions are wrong? Secondly, who is to interpret God's will, and how can we know what God wills in the first place? If Scripture is the answer, what are we to do with diverse interpretations and contradictory passages? If churches are the interpreters of Scripture, what are we to say of their disharmony, the continual multiplication of churches, and their constant bickering and disagreement? How can we adjudicate between the Catholics and the Methodists regarding divorce, for example.

In sum, then, the Divine Will theory is also inadequate since many persons have no contact with His will (agnostics, atheists, and heathens, for example) and even Christians have no clear guide to just what it is that God wills.

As a final note, let me add that because the Divine Will theory is inadequate as a theory does not mean that the great religious teachers are to be completely disregarded. Quite the contrary, most of them were exemplary moral teachers and what they had to say is relevant and should be taken into consideration in making moral decisions. In fact, one of the strongest arguments in favor of the divinity of Jesus is his greatness as a moralist. Nevertheless, the Divine Will theory cannot serve as a general normative theory and the will of God cannot be our ultimate moral standard.

In concluding this section, it should be apparent from the discussion that I personally think that the most adequate forms of normative ethical theory are rule-deontological or rule-teleological. And, since I think that rule-utilitarianism can ultimately be reduced to a rule-deontological theory—with one of the primary rules being teleological—I think that formalism is the most acceptable of normative theories. However, formalism cannot be completely acceptable until we answer some of the meta-ethical questions, such as, what is the ultimate justification of our deontological principles? We will address ourselves to the questions and theories of meta-ethics in the next section.

This account of normative theories is, to repeat, only representative and certainly not complete. The accounts here are also incomplete in the sense that they are mere outlines of the theories. But my purpose in this section was to acquaint the reader with the general types of theory, not to prepare him for his doctoral examination in ethics. I suggest that if maximum benefit is to be derived, the student should do additional reading directly from the original sources. Suggestions for further reading have been included in the notes.

IV

THE JUSTIFICATION OF ETHICAL PRINCIPLES
(Meta-ethics)

It was the point of the last section to discuss the justification
of particular moral judgments. We justify a particular judgment
by appealing to general normative principles. Thus, if I say "your
stealing his money yesterday was wrong," I am prepared to defend my judgment by appealing to the principle, "Stealing is
morally wrong." Thus, we examined the various types of normative theory and their general principles, to see what kind of
principles are used to justify individual judgments.

In this section, we want to go one step further and ask the
question "Can our general principles be justified?" For if these
principles cannot be in any sense justified, then they cannot
serve as justifications for individual judgments. Just as we saw
that there are various theories of justification for normative judgments (deontological, teleological, etc.) in this section we will
see some of the various theories of meta-ethics, each of which
proposes a different answer to our basic question.

Of course, we cannot attempt to justify general normative
principles unless we know what we are saying when we utter
these principles. As a consequence, we must consider another
question at the same time as we are seeking a justification of
general principles. That question is, "What do we mean by our
ethical terms such as 'good,' 'ought,' 'right,' and so forth?" Our
answer to the first question will depend largely on our answer
to the second question. Let us turn now to some of the more
prevalent meta-ethical theories.

EMOTIVISM

Emotivism is a theory which maintains that we do not mean

anything by our ethical terms and that therefore, we cannot even make sense out of the attempt to justify normative principles.[10] We cannot justify nonsense sentences, and if our ethical terms are meaningless, than we utter mere noise when we state a normative principle.

According to emotivism, when we say "You were wrong to steal that car yesterday," we are adding nothing to the factual statement "You stole that car yesterday." We add no more factual information. At most, we are merely venting our emotion of disapproval. It would be exactly the same if we said, "You stole that car yesterday, booh!" and frowned while we said it.

General normative statements fare no better. For the emotivist, to say "murder is wrong" is the same as saying "murder!!!!!!" where the exclamation points indicate, by convention, the strength of your emotion of disapproval. Ethical discourse is meaningless gibberish, pure nonsense. At most it is a cathartic, like sighing, laughing, shouting, or cursing.

Since sighs, laughs, emotions, and feelings do not state anything, they cannot be true or false, reasonable or unreasonable. As a consequence, it makes no sense at all to speak of justifying them, to giving reasons for or against them, to trying to show them true or false. Thus, on the emotivist view, ethical discourse cannot be justified. We were bamboozled into thinking that we were saying something significant when we used ethical discourse only because it resembles factual discourse and only because we really didn't understand what we were doing.

On the surface, emotivism appears to be absurd. To understand it, we must detour for a moment and examine the roots from which it sprang. Emotivism is the ethical theory of logical positivism, a philosophical trend that developed in the early part of this century and was prevalent until the mid-fifties in England and America.

Logical positivism was primarily a reaction to idealist metaphysics of the late 19th century. The positivists contended that the metaphysical utterances of these philosophers were pure nonsense and of no significance at all. To establish their claim, the positivists developed a "criterion of significance" by which we could tell whether or not a sentence was meaningful. To be significant, a statement had to be analytic, that is, refer only to the logical relationship of the terms in the sentence. For example, "A triangle has three angles" is analytic because it states nothing about the world, only about the terms "triangle" and "three angles." An analytic statement is called a "tautology" if it is true and a "contradiction" if it is false. Furthermore, its truth or falsity is logically necessary. Or if the statement is synthetic (a statement which is neither logically true or logically false, e.g. "That wall is green,") there had to be some conceivable state of affairs which would either verify or falsify the statement. The statement "That wall is green" is significant since there is a state of affairs which would verify it, e.g. the wall actually being green, and a state of affairs which would falsify it, e.g. the wall actually being red.

The whole point of the positivist's criterion was to show that scientific statements are meaningful while metaphysical statements are meaningless. In point of fact, the positivists never achieved their goal since they were unable to find a criterion which would admit as meaningful all scientific statements and at the same time exclude all metaphysical statements as meaningless.

But given the positivist criterion of meaning, it is easy to see why they regarded ethical statements as nonsense. They clearly are not analytic; and since we suppose that our ethical statements are saying something about the world, we would not want to try to construct an analytic ethical theory. In addition, it is difficult to suggest what conceivable state of affairs could serve to verify or falsify ethical statements. Thus, they were relegated to the realm of nonsense, with metaphysics, poetry, and babbling.

In addition to the fact that the positivists never produced a criterion that was acceptable to themselves, let alone others, there are several objections against this position. In the first place, we can ask if the emotivist is describing ethical discourse, analyzing it, or recommending a revision in it. If he says he is describing it, we can immediately dismiss his description, because it is clear that his description does not do justice to the whole belief in and practice of justification. We do not think that we are merely venting our feelings when we make a moral judgment. We recognize and in fact, do constantly universalize our ethical judgments, a practice that is not accounted for in his description. There is simply too big a difference between what we are doing when we use ethical discourse and what the emotivist says we are doing to grant much credence to his description.

However, he may say that he is not describing the actual use of ethical discourse, but is analyzing what in fact, we are doing—irrespective of what we think we are doing—when we use ethical terms. Again, I would argue that his analysis is very faulty. He can account for the practices of justification and universalization only by saying that the vast majority of us are under a total illusion as to what we are doing when we think we are making significant ethical statements. And this is a rather audacious presumption to make, especially when he has no internal evidence that we are so illuded. His only evidence is that if our ethical discourse is significant, then his criterion of meaning is wrong. What I am saying here is this: the emotivist's position results not from a careful analysis of ethical discourse but from a prior commitment to his criterion of significance. His conclusion does not result from his analysis, but precedes it. At most, his analysis is circular; at worst, it is no analysis at all, just a reaffirmation of his prior prejudice.

A more clever emotivist may recognize the validity of these criticisms and assert that he is not describing or analyzing actual practice, but recommending that we change our practice to accord with his criterion of significance. Here, he would be saying that although our ethical discourse is not presently a venting of emotion, it ought to be. But other than his feeling better personally, he can adduce no reasons for his revision. We, on the other hand, can give several reasons for rejecting his proposal; and main among them would be that the acceptance of this revision would spell the demise of ethics, something professors of ethics are not wont to allow. Furthermore, we might remind him that his statements about ethics are not analytic, and are not empirically verifiable or falsifiable, and so, by his own criterion, are meaningless.[11]

In short, emotivism is a grossly mistaken description of ethical discourse, is no analysis of it all, and is totally unacceptable as a revisionary plan.

NATURALISM

Naturalism is the meta-ethical view that ethical statements are fact-stating assertions.[12] The naturalists reject, as do many others, the emotivist position for many of the reasons I have given above. If, then, ethical statements are significant and can be justified, the question remains, how can we justify them? The naturalist proposes that we justify them exactly like we justify any fact-stating assertion. The methods of the empirical sciences are quite adequate for justifying—determining the truth or falsity—of ethical discourse.

After agreeing on this point, naturalists have differed among themselves as to what facts are being reported by normative judgments. They all agree that ethical statements are property ascribing statements, but they differ as to what property is being ascribed. One notable naturalist, R. B. Perry, defines "good" as being "an object of favorable interest" and "right" as "being conducive to harmonious happiness."[13]

To determine whether or not something is good, we must know whether or not it is an object of favorable interest, that is, is it desired by someone. Here the techniques of psychology can be used to tell whether or not something is desired. Thus, those same techniques can determine whether or not something is good. As a consequence, for Perry, normative judgments ascribe the natural property of "being desired." This is called a natural property because its presence or absence is determined by ordinary empirical means.

Another form of naturalism is often called *subjectivism,* since on this view, normative statements report the speaker's attitude of approval or disapproval.[14] According to the subjectivist, to say, "Abortion is wrong" is merely to say, "I feel disapproval toward abortion." We have many techniques for determining peoples attitudes, among them personal reports and opinion surveys are predominant. Thus, if it is in fact true that the speaker has an attitude of disapproval toward abortion, then his statement "abortion is wrong" is true.

A variation of subjectivism states that to say, "Abortion is wrong" means that "the majority of the people in my society disapprove of abortion." Of course, the techniques of polling and opinion surveys can determine the attitudes of the majority on any issue.

The most important point to remember about all naturalisms is that ethical terms are defined in terms of natural properties, that is, in terms of properties whose presence or absence can be determined by empirical means.

In addition to what we might call "empirical naturalism" there are theological and metaphysical naturalisms. An example of a theological naturalism would be the divine will theory. Here, "right" is defined as "willed by God." Thus, whether or not some action is right is determined by whether or not God has willed that we do that action. And, the will of God (inasmuch as it can be known at all) is known by theology.

An example of metaphysical naturalism is the natural law theory.[15] On this view, an action is right if it is in accordance with the nature of man. The "nature of man" is a metaphysical notion and so it is ultimately metaphysics which justifies normative judgments. A notable example of how this theory works in practice is in the following. It is the nature of man to procreate in a certain manner. Thus any interference with that manner of procreating is "unnatural", that is, is not in accordance with man's nature. From this, the natural law theory argues that any form of birth control except rhythm is unnatural and consequently, morally wrong.

For all naturalisms, whether they be empirical, theological, or metaphysical, normative judgments are really disguised factual statements. The facts in question may be empirical facts, theological facts, or metaphysical facts. In short, "ought" statements (normative) are translatable into "is" statements (factual statements). And since we have methods of determining the truth and falsity of factual statements (it is clear that we have such methods for empirical statements, although it is debatable whether we have such methods for theological and metaphysical statements) we can easily determine the truth value of ethical statements.

It is not enough to object against naturalism on the grounds that "ought" statements are not translatable into "is" statements. That would be begging the question, since this is the central thesis of naturalism. We must show why this is not possible.

One very famous argument against naturalism was advanced by G. E. Moore, who first formulated the "open question test."[16] Suppose we define the normative term "good" in terms of some natural property P (e.g. object of favorable interest, object of approval, etc.). Then we ask, would it make sense to ask, "X has P, but is it good." If the question does make sense, the definition is incorrect. Another way of putting it is that if the definition is correct, the question is not significant, for it would be asking, "X has P, but does it have P?" And this is clearly not a significant question. Moore's contention is that no naturalistic definition can pass the "open question test." Thus, he concluded that to define ethical words in terms of natural properties, whether they be empirical, theological, or metaphysical properties, is a fallacy, and he dubbed this the "naturalistic fallacy."

Though on the surface this test of ethical definitions appears quite strong—and was accepted for many years—it is somewhat defective. First, a naturalist may answer that ethical terms have several meanings and many connotations. Thus, even though their definition may be correct for one meaning, it may be significant to ask Moore's question, without realizing that the question is significant only for one of the other meanings not defined in that particular definition. This is clearly the case with such terms as "ought" and "good" and "right" all of which have many meanings.

A second objection is that we have no clear criteria for significance, thus, a naturalist may deny that we can ask Moore's question significantly. Perry, for example, may insist that it is not significant to ask, "X is an object of favorable interest, but is it good?" whereas Moore would argue that the question is significant. If both hold their ground, how can we adjudicate between them. Certainly, in some cases the significance or lack of it with regard to Moore's question is obvious. But in many cases it is not; and, unfortunately, in the case of ethical terms, it is not obvious whether or not the question is significant.

A third possible tack the naturalist could take against Moore, and one actually taken by Perry, is to admit that his definitions are not accurate accounts of what we mean by our ethical words in ordinary language, but propose that his definitions be adopted for future use. In sum, then, the "open question" test may indicate that something is amiss with the naturalist's definitions, but it is clearly insufficient to prove the point. Moore's argument is simply insufficient to reject naturalism on this ground alone.

There are at least two additional arguments, both of which are much stronger than the first, to show that naturalism is mistaken. The naturalist is essentially asserting that we have two alternative vocabularies with which to state facts. The usual fact-stating vocabulary and the vocabulary of ethics. This suggestion strikes me as very queer. For why do we need two fact-stating languages? Couldn't we clear up many of our difficulties—certainly all of the difficulties in ethics—by restricting ourselves to the usual fact-stating language? In addition, it seems clear that when we use ethical discourse it is not merely to state facts. If it were, then we could quite adequately do without this special way (ethical discourse) of stating facts. When we say "It is morally right to be truthful," we are not merely stating some fact about truth telling, like it is "conducive to harmonious happiness." We are recommending it, prescribing it, perhaps commanding it, expressing our favorable attitude toward it, and reaffirming our intention to censure anyone who lies. It seems rather presumptuous to me that someone would argue that although I think I am doing all or some of these things when I use ethical words, in reality I am only stating facts. For, if I want to state facts, I know how to make clear factual statements. Thus, when I use moral language, I am intending to do more. And if the naturalist wants to say that most of us have been totally mistaken about what we are doing when we use normative discourse, he needs to present us with much more evidence then he can produce.

The second of the stronger objections is this. When the

naturalist offers us his definition of a moral term, he is implicitly asking us to accept the corresponding principle. F. C. Sharp has defined "right" as "desired when looked at from an impersonal point of view."[17] If we accept this definition, we are also accepting the normative principle, "One ought always do that action which is desired when looked at from an impersonal point of view."

To use a definition as the justification of a moral principle—and this is what the naturalists are doing—presents us with two problems. In the first place, we have just displaced our search for a justification from the principle to the definition. If the principle is justified by the definition, how do we then justify the definition? And if the definition has no justification, then neither does the allied principle. Secondly, since the acceptance of a definition implies the acceptance of a moral principle, we cannot justify the principle by the definition. That would clearly be circular.

Let me summarize my objections to naturalism: It suggests that we have two alternative fact-stating languages. It argues that, in spite of what we think we are doing, we are merely stating facts when we use ethical discourse. It proposes that we justify moral principles by definitions, a procedure that is either circular or self-defeating since justifying a reportive definition is notoriously difficult when there is substantial disagreement over the meaning of the term being defined. Thus naturalism is, in my view, correct in its contention that normative judgments can be justified, but mistaken in its view of how this is accomplished.

INTUITIONISM

Intuitionism is the meta-ethical view that our basic moral principles are self-evident and we come to know they are true by intuition or "rational insight."[18] If this is the case, they are justified in themselves and cannot be justified by showing their logical connection with any empirical, theological or metaphysical facts. Most formalists have been intuitionists with regard to the justification of the basic moral principles.

With regard to the meaning of ethical words, the intuitionists have argued that ethical words ascribe ethical properties. Such properties are not natural properties, not theological or metaphysical properties. They are simple, unanalyzable non-natural properties.[19] This means that they cannot be defined in terms of any other words because they are simple concepts like "yellow." For the same reason, i.e. their simplicity, they cannot be analyzed into their components. They are non-natural because we do not determine their absence or presence by empirical means, that is, by sense experience. Yellow is a simple, unanalyzable, natural property because it is empirically known. "Good," however, refers to a non-natural property since we cannot know whether or not something is good by any form of sense experience. We know non-natural properties only by intuition.

Intuitionism, in my opinion, is fatally defective on both counts; in its view of justification, and in its view of the meaning of ethical words. Taking the latter first, I think that the whole notion of non-natural properties is simply too vague to withstand examination. Intuitionists have so far provided us only with a negative description of non-natural properties. So far, we are still waiting for a positive description of these alleged properties. It seems to me that when someone asserts that there are such unusual properties (and as far as I know, only ethical properties are non-natural) he is obliged to give us a better description of them and more evidence of their existence.

Secondly, to justify basic moral principles by intuition is no justification at all. This is clearly the case since we have no criteria for distinguishing between veridical intuitions and false intuitions. This is critical in ethics since we often have cases of conflicting basic moral principles. Both cannot be true. Both parties argue that their principle is intuitively self-evident. How can we adjudicate? By what criteria would we decide that one intuition is veridical, while the other is false? Furthermore, to admit that we have no criteria for adjudicating intuitions, and to rest one's normative principles on intuition, is tantamount to reducing ethics to vulgar opinion.

I am sympathetic with the guiding motives of the intuitionist position. Naturalism is inadequate. Our moral principles do appear to us as self-evident. Ethical terms do not ascribe natural properties to actions, persons, motives, and intentions. But I disagree with their solution to the problem. Although our basic principles do often appear self-evident, absolutely immoral principles have appeared self-evident to their subscribers. Thus, if ethics requires us to justify our principles, intuition cannot be enough. Secondly, ethical terms do not ascribe non-natural properties either. For, the notion of a non-natural property is very suspect and it is questionable whether or not ethical statements ascribe any kind of properties. Ethical statements do not state facts, natural or non-natural.

NONCOGNITIVISM

Noncognitivism is a recent meta-ethical theory which agrees with emotivism in one important respect and agrees with naturalism and intuitionism in another important respect.[20] Noncognitivism and emotivism are in agreement that ethical statements do not state facts or ascribe any kind of properties to persons or actions. There can be no inductive or deductive proof of the correctness of an ethical statement. There are no natural or non-natural facts which can be used as evidence for or against ethical statements. On this point, the noncognitivist rejects the meta-ethical claims of both the naturalists and the intuitionists.

However, on the matter of universalizing and justifying ethical judgments, the noncognitivist sides with the naturalist and the intuitionist against the emotivist. It will be remembered that for the emotivist, ethical judgments cannot be universalized because they are the expression of a personal feeling or emotion. They cannot be justified since it is not clear how we go about justifying an emotion or feeling. On the emotivist's view, it just doesn't make sense to speak of justifying an expression or venting of emotion.

The noncognitivist rejects the emotivist view on two counts. First, ethical statements are not expressions or ventings of emotion. Secondly, ethical statements are universalizable and justifiable, albeit in a slightly modified sense.

According to the noncognitvists, ethical terms and statements are used for many purposes. "They are used to express tastes and preferences, to express decisions and choices, to evaluate, to advise, admonish, warn, persuade and disuade, to praise, encourage and reprove. . . ."[21] It is not, however, noncognitivists insist, the function of ethical terms and statements to state facts or ascribe properties.

Sometimes ethical judgments do convey information. For example, the judgment "You were wrong to steal that student's books" implies the factual statement, "You stole that student's books." Although often factual information is conveyed by ethical statements, their primary function is different.

According to one of the more popular views, that of R. M. Hare, the function of ethical statements is to *prescribe*.[22] And, prescription is something done by commands, requests, cookbooks, and pharamceutical prescriptions. Prescriptions tell us what to do. Factual statements tell us what is the case. Prescriptions are directive, while factual statements are informative. General ethical statements are a distinct kind of prescription, a universal prescription.

"You ought not to smoke" is a particular judgment which, according to Hare, is equivalent to the prescription, "You, don't smoke!" The general statement, "Smoking is morally wrong," is equivalent to the prescription, "No smoking by anyone at any time in any place!"

This is how Hare accounts for the universalizability of ethical judgments. In the first place, the prescriptions are reflexive. When I prescribe no smoking by anyone, I include myself. Secondly, prescriptions are universalized not only to all probable situations and persons, but to all logically possible situations or persons. This gets around the following problem. Suppose I am a rich, powerful, secure slave owner. I could then universalize my prescription, "A slave owner may treat his slaves any way he pleases," knowing full well that I will never be a slave. However, if I must universalize them to include all logically possible cases, I would be loathe to offer my prescription since it is logically possible that I could be a slave, and in this circumstance I would hardly agree with the prescription. Thus, prescriptions are universalizable and must be extended to every logically possible contingency to be valid.

Hare has also recognized the need and the legitimacy of justifying ethical statements. He says that prescriptions can be justified in the sense of giving reasons for them. For example, when someone asks his physician why he has prescribed this particular drug, he may say, "It is the best drug I know of for clearing up your condition." When I make the prescription, "No smoking" I am ready to support it with reasons, like, "This is a crowded room with no ventilation," or, "Smoking has been shown to be deleterious to health."

The noncognitivists are correct, in my view, on two points. First, ethical statements are not fact-stating, informative statements. Ethical statements perform many functions, sometimes one function, sometimes another, and sometimes several at once. Secondly, they are correct in reaffirming that ethical judgments must be universalized and can be justified.

Nevertheless, I think that they are mistaken on several points. First, I doubt that the primary function of ethical judgments is to prescribe. For prescriptions are justified primarily by the authority of the person making the prescription. It is only when this authority is in doubt or when the prescriptions go beyond the scope of the authority that we insist on reasons to justify them. Secondly, the noncognitivist does not carry far enough the procedure of justification. According to the theory, we have justified a prescription when we have given reasons in terms of generally accepted norms. For example, the health reason for not smoking would be no reason at all if we did not generally value health, disvalue threats to health, and consider long life a positive good.

In short, justification cannot stop at the offering of public reasons (in contradistinction to purely personal reasons) in support of our ethical judgments. For, if we do not attempt to go further, we are saddled with the ultimate relativism of attitudes that offers no hope of adjudication. An example here may be enlightening. Jones asserts "Abortions are morally permissible as long as they are performed by a competent physician." He supports his prescription with the following reasons. "There are already too many abortions performed each year by incompetent quacks who often fatally injure the women involved." "Very often the unwanted pregnancy is the result of criminal assault or rape."

Smith, however, asserts, "Abortion is morally wrong." He supports his claim thus: "Abortion is the killing of a human person." "To legalize abortion simply because illegal abortions are taking place is similar to legalizing burglary because illegal thefts are taking place. What is needed is enforcement, not a change in the law."

Both have made universalizable moral judgments. Both have supported them with publically acceptable reasons. Here the noncognitivist would have to say that both judgments are justified, even though they are contradictory. This I find hard to accept, especially without attempting to carry the process of justification to a higher level. In the following, I will present my meta-ethical theory in which I will show how justification can be carried to its logical limit.

AXIOM THEORY

Axiom theory is the meta-ethical view that basic normative principles should be considered as axioms in an axiom system and that this system is analogous to axiom systems in science.[23] This means that an ethical theory will formally ressemble any axiom system and that normative theories can be evaluated by many of the same criteria that we use when evaluating axiom systems in science.

As was evident from my treatment of formalism, I think that it is the most adequate normative theory. However, its one major drawback was its dependence on intuitionism for its meta-ethical justification. The problem with intuitionism is the reliance on "rational insight," "intuition," or the "self-evidence" of basic normative principles. My position is that formalism is acceptable if normative principles are treated as axioms and are justified if and when the whole theory in which they play a part is justified.

There may be any number of formal normative principles accepted as axioms. All that is required is that the axiom theory satisfy the criteria of consistency, universalizability, relative completeness and parsimony.

A particular normative judgment is justified by reference to a rule. The rule, in turn, is justified by reference to a basic principle. Finally, the basic principle is justified by reference to the whole axiom system in which it is an axiom. In the case of competing axiom systems, that whole system is preferable which best satisfies the criteria outlined above.

In the case of only one axiom system, that system is itself justified if there are no competing axiom systems. In other words, if there is only one consistent and universalizable axiom system, it is better than none at all. It is generally supposed, however, that the disagreements in ethical matters indicates that there is more than one axiom system. If this is the case, we must adjudicate among them and we have criteria to guide this evaluation.

It is important to note, however, that only another axiom system can displace an axiom system. A conflicting particular judgment or even a conflicting principle is insufficient to lead us to reject an axiom system. Only another fully developed system is sufficient for this.

If, in terms of the criteria, there are no reasons for preferring one axiom system to another, then both are equally acceptable (and the particular moral judgments deducible from them are equally justified), even in the unlikely case that an axiom of one system contradicts an axiom of the other. This means that if we have no reason for choosing one system over another, both are tentatively allowed to stand side by side until one can be shown to better fulfill our criteria than the other.

It follows from this that once acceptable systems may become unacceptable in the face of a new competing axiom system which better fulfills the criteria. This means that no system is sacrosanct and forever valid, but that ethical theories, like all other theories, are subject to constant review and continual modification, especially in the light of new and different circumstances.

Perhaps if we examine in some detail the analogy of axiom theory in ethics with axiom theory in science, this position will become clearer. A particular empirical judgment is ultimately justified (accepted as true or as confirmed) only in reference to the total theory in which it plays a part. No empirical statement is either true or false, justified or unjustified in isolation.

Secondly, we no longer accept the most basic axioms of scientific theories as self-evident or as absolutely true. They are accepted as axioms only and are justified only if the whole theory is found to be superior to any competing theory on the basis of the criteria.

Thirdly, a scientific theory is refuted only by a rival theory. No isolated empirical statement is sufficient to lead to the rejection of a theory. A case in point here are the phenomena of parapsychology. It is increasingly difficult to discredit as frauds or illusions all of the reports of such things as clairvoyance, psychokinesis, etc. Yet, we prefer to reject these isolated empirical reports rather than reject our current physical theories of perception and causality. As a consequence, we will continue to reject parapsychology, until it is able to organize the empirical phenomena into a coherent, explanatory system. At this time, we will have two competing theories and can accept or reject one or the other or both on the basis of our criteria for evaluating theories.

If we can find no reason in terms of the criteria for preferring one scientific theory to another, either is acceptable until we do have such reasons, or until we have a third and superior theory. An example here is the acceptability of both the wave and the particle theories of light. Until we have reason to prefer one over the other, scientists are justified in choosing either one. In this case, the photon theory may be preferable, but until it is shown to be superior, we may have three acceptable rival theories. All are under constant review and subject to rejection at a later date when reasons can be advanced for accepting one and rejecting the others. Until then each must be considered as justified and acceptable.

The history of science is full of cases where once acceptable theories have become unacceptable in the face of a competing theory which better fulfills the criteria. Some examples here could be the rejection of the impetus theory of motion in favor of the inertial theory; circular inertia in favor of rectilinear inertia;

Newtonian physics in favor of Einsteinian physics, and phlogiston theory in favor of the oxidation theory.

Let us now take a look at some of the consequences of axiom theory in ethics. First, and most importantly in my mind, a deontological normative theory such as formalism no longer depends on intuitionism. Secondly, we can make sense of the justification of basic normative principles without reliance on naturalism, the absurd abandonment of justification by emotivism, or the weak justification procedure of noncognitivism.

Thirdly, an ethical judgment is no longer construed as a fact-stating sentence, nor does it ascribe natural or non-natural properties. Furthermore, we can recognize that ethical statements do perform many functions as the noncognitivist has correctly argued. Finally, we can see that ethical statements are validity-claim making statements. They claim that they are part of an acceptable axiom system in which the axioms are the most basic answers to the question, "What shall I do" or "How shall I live." An essential part of this claim is that they will stand up under public scrutiny and fulfill, either directly or indirectly, publicly accepted criteria for validity.

In this sense, ethical statements are practical and directive. But this does not mean they are imperatives or commands, expressions of attitude, exhortations, etc. They are, in my view, *sui generis*, although they bear a family resemblance to commands, expressions of attitude, exhortations, etc., and are often used to perform the same functions as these types of linguistic expression.

The root of past error has been the failure to see that ethical terms and statements are *sui generis* and are not reducible to other forms of linguistic expression.

In the fourth place, although axiom theory in ethics is formally analogous to axiom theory in science, no ethical axiom system may be derived from any scientific axiom system and vice versa. This is because the axioms themselves are of an essentially different nature.

Fifthly, it is possible to justify particular moral judgments to the same extent it is possible to justify particular empirical judgments. However, the exact technique is slightly different. Thus, it makes sense to differentiate among verifying logical statements, confirming empirical statements, and assaying ethical statements. They are different types of statements and must be evaluated in different ways. The historical error here was the tendency to conclude that ethical statements cannot be rationally evaluated at all since they can not be confirmed or verified. Would we want to say that empirical statements are unjustifiable because they are not verifiable?

Sixthly, an important consequence of this theory is that much disagreement in ethics can be settled by formalizing the principles and theories involved. Very often disagreement arises because one of the disputants is relying on a basic normative principle which would obviously be seen as unacceptable if it were formalized into an axiom in a system.

Since we have a technique for adjucating between competing axiom systems, we do not need to admit that we can reason no further and that the dispute ultimately rests on a matter of choice until we have determined that both of the competing theories equally fulfill all of our criteria. There is no reason, I think, to abandon reason before this point. If ethics is to be rational at all, it is not until this point that we can say that the disagreement ultimately rests on attitudes.

At this point I think I am justified in advancing the following prediction. It will be rare that even competing theories will justify contradictory particular moral judgments in any specific case. This means that the consequences of allowing the possibility of two equally acceptable competing theories will not be as serious as it is often thought.

One last note worth mentioning is that axiom systems could contain either teleological or deontological principles or a combination of both. I personally think that all of the axioms would be deontological since I have argued that even teleological principles are ultimately deontological in nature.

Let me now offer a rough sketch of a possible axiom system in ethics so we may see what it would look like when worked out.

> *Axiom 1.* Each person ought to protect and enhance his own freedom and the freedom of others. For, without freedom, no

ethical choice, no moral or immoral action, and the whole concept and institution of ethics, is impossible. Where the exercise of my freedom conflicts with the exercise of others' freedom, each man shall count as one and the majority shall exercise its freedom.

Axiom 2. No one should injure, harm, or kill another person. Furthermore, there is an obligation of reparation for any intentional or accidental harm done another.

Axiom 3. There is an obligation of justice.

Axiom 4. There is an obligation of fidelity.

Axiom 5. There is an obligation to help and render assistance to anyone in need. The more dire the need, the stronger is the obligation.

Note: The axioms are ranked in hierarchial order and in cases of conflict, the axiom with the lower number takes precedence.

Rules: (Partial list)

Rule 1. No one ought to keep another person as a slave or in involuntary servitude. (Axiom 1)

Rule 2. No one should surrender his own personal freedom by the use of debilitrating or addictive drugs. (Axiom 1)

Rule 3. Economic servitude is a serious form of slavery and should not be encouraged or permitted. (Axiom 1)

Rule 4. No one should act in such a way that a foreseeable consequence of his action is the injury or death of another person, even if this injury or death is not intended, e.g. driving a car while drunk. (Axiom 2)

Rule 5. Injuring a person's reputation or causing him to loose self-esteem and self-confidence is a real and serious form of harm. Mental pain is as reprehensible as physical pain. (Axiom 2)

Rule 6. One should always encourage and never prevent anyone's receiving equal treatment before the law or in court litigation. (Axiom 2, Axiom 3)

Rule 7. Property is not an absolute right. Material goods should be distributed in such a way that no human being is denied the basic minimum material necessities of life as far as this is possible. (Axiom 3, Axiom 5)

Rule 8. One should always be truthful to those who have a right to know the truth, and honor and respect his promises where it is within his reasonable power to do so. (Axiom 4)

Certainly there may be many other moral rules justified by

these axioms. Where rules conflict, the rule justified by the lowest numbered axiom takes precedence.

It is obvious, I think, how these rules can be used to justify particular moral judgments in specific cases.

Clearly, it is possible to have another axiom system with different axioms. However, we cannot adjudicate between this axiom system and any other until the other is presented in at least as much detail as this one. At that time, we may decide on the basis of our criteria that the competing system is preferable. Thus, it is encumbent on those who would not accept my axiom system to present an alternative.

V

A NOTE ON VALUES

Up to this point, our attention has been directed toward the development of a theory of obligation. We have been trying to work out a theory of action and a means of evaluating the rightness or wrongness of actions. In order to complete our theory of obligation, however, we must look briefly at the question of values. It is not within the purview of this book to develop here a complete account of values. Thus, the following should be taken as remarks on value rather than an attempt to do full justice to the topic of values.

When we were discussing the normative theory of utilitarianism, we left, for the time being, the term "utility" uninterpreted. On that theory, an action was right if it produced a maximum of "utility" as compared with any alternative action. In the past, "utility" has received diverse interpretations. For Bentham, "utility" meant the balance of pleasure over pain. For others, it has meant "well-being," that is, pleasure, joy, intelligence, knowledge, contentment, satisfaction, etc., rather than pain, sorrow, ignorance, error, anxiety, and dissatisfaction.

Other theories, such as formalism or my version of an axiom theory, incorporate values into the principles of obligation. Such values as freedom, life, justice, and love were seen as integral parts of the normative principles themselves. And, it is my belief that no normative theory is adequate until values have been included in the principles of obligation. In the case of utilitarianism, we would have to wait for a complete appraisal until "utility" is interpreted—which, in fact, it is in almost all accounts of it. The reason that value and obligation are separated at all is for the sake of clarity. For example, the utilitarians might leave "utility" uninterpreted for the sake of arguing the merits of a teleological theory over a deontological theory. Or an act-utilitarian may want to abstract from values while presenting his case against a rule-utilitarian.

There are several types of value. Nonmoral value is the assessment of worth of nonmoral actions and objects. This is the type of value that we place on cars, wine, beef steak, and other material objects. We correctly call these things "good" or "bad" but not "morally good" or "morally evil." Nonmoral value is usually attributed to something as a result of our estimate of how well it accomplishes its purpose or of how high a quality the object is compared to like objects.

I am thinking here of two examples. The first is our calling a particular automobile "good" on the basis of how well it performs its function. When something has several functions, it can be correctly called "good" with reference to one function and "bad" with reference to another. A small sports car is certainly "good" with respect to the function of high speed touring for a limited number of persons over narrow, winding roads. American passenger cars are not well suited for this function. On the other hand, for the function of carrying six people at high speed on Interstate highways in a country where gasoline is relatively cheap, an American car would be "good", or "better than a sports car." The second example I have in mind is the grading of wine. Wine is evaluated in direct comparison to other wines. In both of these examples, we have criteria for such an evaluation. In evaluating sports cars, we look to acceleration, suspension geometry, braking, handling, and responsiveness. For American cars, we want comfort, power, luxury, and reliability. Wine is evaluated on the basis of dryness, aroma, body, stability, color, and taste.

The point that I am trying to establish here is that we have criteria for evaluating nonmoral good and it is not solely a matter of attitude or taste. For example, I may admit that a Datsun fulfills all of the criteria for a good sports car, but insist that I do not like it. I can give reasons why it is a good car, but I often am unable to say why I do not like it. This applies equally well to wine. By every criteria, Nuits St. Georges, 1959, is an excellent wine. But I do not like it. I don't know why, "It just tastes funny."

A second kind of value is called "instrumental" value. Something is instrumentally good if it is an effective means to something else. A Cadillac may not be a good car, but it is certainly instrumentally good if the goal is to achieve status in a suburban neighborhood. Effectiveness is the criteria by which we judge instrumental goods.

The third kind of value is called "intrinsic" value. Here something is said to be good in itself, rather than as a means to something else. Happiness and pleasure are examples of intrinsic goods. Notice that instrumental and intrinsic goods can be either moral or nonmoral. The observant reader has by now seen how complex the whole matter of value really is. For it is not always clear what sense of "good" we are using when we call something good and many things are good in one sense and not good in another.

When we apply "good" to persons, we are usually referring to character traits, that is, to dispositions to act in a certain way. Examples of character traits are wisdom, temperance, courage, humility, charity, prudence, and so forth. With respect to these traits, there can be at least three kinds of theory, corresponding roughly to the types of normative theory. "Trait-egoism" is the view that certain character traits should be fostered because they are most conducive to our welfare. "Trait-teleological" theories urge that traits be cultivated because they are conducive to the greatest general good. Lastly, "Trait-deontological" theories argue that certain traits should be encouraged because they are intrinsically good, irrespective of any nonmoral value they may promote. All three types of theory may agree on what traits should be fostered, but differ on the reasons why.

It has been urged by at least one writer that ethics be considered solely as the cultivation of certain moral character traits.[24] Thus, that man is good who is of a morally good character. Furthermore, we can judge actions as right or wrong depending on whether or not they issue from a morally good disposition. According to this theory, we should be primarily concerned with cultivating good characters and only secondarily concerned with the principles of obligation. This view has much to recommend it and was very popular among the ancient Greeks, however, I find two difficulties with it.

First, we can determine a person's character only by his actions. Thus, we say a man is prudent as a result of seeing him act prudently. It is his courageous actions which lead us to call him courageous. Thus, it seems to me that the evaluation of actions is logically prior to the evaluation of persons. Secondly, in some instances a person's actions may be contrary to his dispositions. A "good" man may commit moral evil. We say, "He acted out of character." This means that his action is inconsistent with what we know of his character traits. And this, I submit, means that the action is inconsistent with the trend of his past actions. For both of these reasons, I think that an evaluation of action, a theory of obligation, is a prerequisite for an evaluation of character.

The interrelation of obligation and value should be readily apparent. It is not enough to say that actions should produce the maximum net expectable utility until and unless we know what nonmoral value is being produced. A man is said to be morally good because he performs right actions and our evaluation of actions as right depends on an assessment of values.

The next to the last point that I want to make is this. Values are in some sense relative. I do not think that we can establish an hierarchy of values or "a most fundamental intrinsic good." This is especially true with nonmoral values. Some people value leisure, while others challenge. Some value a long, secure and easy life, while others want an exciting, challenging life, even if it is short. I know no way of adjudicating between such disagreements of value. However, values are universal in some sense. All of us are similar in many respects; all of us have similar needs and similar desires. In this sense, I think there is not really as much fundamental disagreement about values as is often thought. All of us would agree, I think, that happiness, pleasure, joy, contentment, intelligence, and well-being are intrinsic values. We might not agree that they should be ranked in a certain order, but it is my contention that order is not terribly important for ethics. Values seem to be rooted in our nature and our condition to an extent sufficient for a rational ethics.[25]

Lastly, we might with profit distinguish between "a good life" and "the good life." A person is said to have led a good life if he was a moral man, that is, if his actions were morally right. On this we can all agree. We may not agree, however, on what constitutes the good life. For a Jewish patriarch, it was having his children, grandchildren, and great-grandchildren around him; prosperity, good health, and freedom from worry. For an artist or poet, the good life is time to devote to his art, an intensity of experience, and success in communicating those experiences. To a middle-class American, the good life is a house in suburbia, two cars, a good job with an adequate income, two children who do well in school, and barbeques in the back yard on weekends. In sum, "the good life" is relative to the attitudes and desire of the individual; "a good life" is not.

VI

A NOTE ON FREEDOM

Although it is not possible here to fully discuss the issue of human freedom, it is not possible to adequately treat ethics without saying something about it. For, if man is not free then morality is impossible and ethics is the absurd study of the impossible. It simply does not make sense to say that someone is morally wrong for not doing something he could not do. Likewise, it is senseless to praise someone for doing something he had to do.

In the history of philosophy, the discussion of ethics and freedom has taken at least three general forms. First, it has been argued, by Holbach, Paul Edwards, John Hospers, and others, that man is not free.[26] Rather, he is just as determined by physical or psychological laws as any other natural object in the universe. From this, the hard determinists conclude that ethics and all talk of morality are meaningless.

Others, like Hume, Mill, and Schlick argue that man is determined in a certain sense, but that this did not mean he was not free in a sense, sufficient for morality.[27] Their arguments took at least two forms. First, Mill argued that all of our actions are caused, but that we are able to modify the causes of those actions, and thus had some indirect control over them. Secondly, Schlick and many contemporary analytic philosophers distinguish between "freedom of choice" and "freedom of action." They denied that we have the first, since our choices are determined by antecedant factors over which we have no control, but that we do have freedom of action, that is, we can usually put into action those choices, and that this is all that is needed to make sense of responsibility.

A third position was to deny that the soft determinists had solved the problem and to insist that responsibility requires both freedom of choice and freedom of action. I have argued at length elsewhere that the soft determinist position is inconsistent and that hard determinism is in error.[28] Both deterministic positions, on my view, result from the illegitimate transfer of the concept of causality from the naturalistic languages (e.g. the languages of the natural sciences) to a phenomenological context (e.g. languages expressing the experiences of persons). That such a transfer is a fallacy is easy to show. It commits the determinists to the unusual position that nonmaterial things such as ideas, desires, motives, and purposes cause physical actions, a position that is quite inconsistent with everything that we know about causality and physical events. The fallacy committed by the determinists is very much akin to the fallacy of anthropomorphism, that is attributing the properties of persons to impersonal beings. We have recognized for many years that anthropomorphism has no place in science. But only recently has it become clear that "scientism" has no place in our discourse about the experiences of persons.

It is just as much of a logical mistake to describe a person's choice as the effect of certain antecedant causes as it is to explain acceleration of a falling body as, "a falling object moves faster as it nears the earth because it becomes more 'jubilant' as it approaches its natural resting place."

That such a transfer of a scientific concept (i.e. "causality") to a nonscientific context is unnecessary follows from the fact that we have a very adequate language in which to describe the experiences of persons; and this language does not have the naturalistic concept of causality, and it is not deterministic.

My position can be summarized as follows. Freedom is at the core of our experiences of ourselves and our actions. No plausible argument has been advanced to lead us to doubt this. As a consequence, there is no reason to believe that we are not free in the full sense required by ethics.

Peter Caws suggests that at very least, we can maintain freedom as an unrefuted hypothesis.[29] He is quite correct in his arguments that freedom can never strictly be proven. But, according to him, it has not been disproven and so we are justified in maintaining it as an hypothesis. On this point, I think he is mistaken. Freedom itself is not and cannot be a hypothesis since it is an absolute prerequisite for the whole activity of theorizing. Professor Charles Landesman has clearly made this point when he says the following:

> However, any rigidly deterministic theory of the creation and discovery of philosophical ideas by intellectuals whose business it is to philosophize, any theory to the effect that language, or economics, or culture, or the unconscious necessitates certain ideas is bound to be inadequate because it denies a condition under which any theory, including itself, can be made acceptable; namely, that the theory is acceptable because of or on the basis of reasons, or evidence, or argument.[30]

As a consequence, I would argue that freedom is not an unrefuted hypothesis, but the prerequisite for any kind of hypotheses at all. Freedom is clearly a prerequisite for any intellectual activity. This, I think, does in no way endanger ethics or science or any other intellectual activity since freedom is at the very core of all of our experiences. Paul Ricoeur has argued, and correctly, I think, that even those situations in which we experience ourselves as not being free, as being determined, are intelligible only in reference to freedom.[31] We can understand them only in relation to the freedom which they deny.

VII

CONCLUSION

If I have been at all successful in this chapter, the reader will now have a cursory view of ethical theory and the logic of ethical argumentation. Most importantly, he will see the relation between ethical theory and normative judgments. Thus, if the normative theory which I have offered is found wanting, he will see the necessity of replacing it with a superior theory.

If, then, we understand the structure of ethical arguments and the relations between particular moral judgments, moral rules, basic principles, and general theories, we are in a position to apply this knowledge to specific cases of ethical questions arising in science. And, this is the primary goal of our whole effort.

REFERENCES

1. On situation ethics, see Barnes, Hazel: *Existentialist Ethics.* Knopf, New York, 1967, and Fletcher, Joseph: *Situation Ethics, The New Morality.* Westminster Press, Philadelphia, 1966.
2. Frankena, William: *Ethics.* Prentice-Hall, Englewood Cliffs, 1963, p. 22.
3. See Warnock, M.: *Existentialist Ethics.* Macmillan, London, 1967.
4. Ross, W. D.: *The Right and the Good.* Clarendon, Oxford, 1930, p. 21.
5. *Ibid.,* p. 41.
6. *Ibid.*
7. Ross, of course, does not claim his list of rules is complete. See Ross, *op. cit.,* pp. 20, 23.
8. See Smart, J. J. C.: *An Outline of a System of Utilitarian Ethics.* Melbourne University Press, Australia, 1961. See also Brandt, Richard: *Ethical Theory.* Prentice-Hall, Englewood Cliffs, 1959, pp. 380-391, and Frankena, *op. cit.,* p. 32.
9. On rule-utilitarianism, see Brandt, *op. cit.,* pp. 400-405; Frankena, *op. cit.,* pp. 33-35, and Rawls, J.: Two concepts of rules. *Philisophical Review,* LXIV: 3-32, 1955.
10. Ayer, A. J.: *Language, Truth and Logic.* Victor Gollancz, London, 1948; Stevenson, C. L.: *Ethics and Language.* Yale University Press, New Haven, 1944, and Brandt, *op. cit.,* pp. 205-221.

11. On this point, see Hemple, Carle: Problems and changes in the empiricist criterion of meaning. *Revue Internationale de Philosophie,* 1950.

12. For more on naturalism, see Brandt, *op. cit.,* pp. 151-183, and Perry, R. B.: *Realms of Value.* Harvard University Press, Cambridge, 1954.

13. See Perry, *op. cit.,* pp. 3, 107, 109.

14. See Westermarck, E.: *Ethical Relativity.* Harcourt, Brace Co., New York, 1932.

15. For a complete exposition of natural law theory, see Fagothey, Austin: *Right and Reason.* Mosby, St. Louis, 1963. See especially Chapters 8-11.

16. See Moore, G. E.: *Principia Ethica.* Cambridge University Press, Cambridge, 1929.

17. Sharp, F. C.: *Ethics.* The Century Co., New York, 1928, pp. 109, 410-411. See also Brandt, *op. cit.,* pp. 259-262.

18. See Ross, *op. cit.,* and Ewing, A. C.: *Ethics.* English Universities Press, London, 1953.

19. See Moore, *op. cit.*

20. On noncognitivism, see Hare, R. M.: *The Language of Morals.* Clarendon Press, Oxford, 1952. Also see Nowell-Smith, P.: *Ethics.* Penguin Books, Baltimore, 1954, and Warnock, M.: *Existentialist Ethics.* St. Martin's Press, New York, 1967.

21. Nowell-Smith, *op. cit.,* p. 98.

22. See Hare, *op. cit.*

23. I have been greatly influenced in my views here by this book. Caws, Peter: *Science and the Theory of Value.* Random House, New York, 1967.

24. Stephen, Leslie: *The Science of Ethics.* G. P. Putnam's Sons, New York, 1882.

25. See Rogers, Carl: Toward a modern approach to values: the valuing process in the mature person. *Journal of Abnormal and Social Psychology,* 68:160-167, 1964.

26. See Baron d'Holbach, Paul Henri: *The System of Nature.* E. Truelove, London, 1863 (First published 1770); Edwards, Paul: Hard and soft determinism. In Hook, Sidney (Ed.): *Determinism and Freedom in the Age of Modern Science.* New York University Press, New York, 1958, and Hospers, John: What means this freedom. In Hook, *op cit.*

27. Mill, J. S.: *A System of Logic.* 1843. Especially Chapter 2, Book VI, and Schlick, Moritz: *Problems of Ethics,* David Rynin (trans.). Prentice-Hall, New York, 1939.

28. See the author's doctoral dissertation, *Freedom and Determinism*. University of Kansas, Lawrence, 1967.

29. See Caws, *op. cit.*, Chapter VI. See especially pp. 87-91.

30. Landesman, Charles: Does language embody a philosophical point of view? *Review of Metaphysics*, June, 1961, p. 619.

31. Ricoeur, Paul: *Freedom and Nature, The Voluntary and the Involuntary*, Erazim Kohac (trans.). Northwestern University Press, Evanston, 1966.

Part Two

A Casebook of Ethical Issues
in Scientific Research

VIII

PURPOSE AND USE OF THE CASEBOOK

PURPOSE

Since it has been the whole orientation of this book not only to acquaint the reader with the rudiments of ethics but also to give him practical experience in dealing with the kinds of ethical problems he will eventually face, it would be seriously remiss to stop our studies at this point. It seems to me that it is not enough to have a theoretical knowledge of ethics. If it is to be of any importance to practicing scientists, the application of ethical theory must also be studied. It is to this end that this casebook has been included.

The purpose of the casebook is to give students and other interested persons an opportunity to (a) become familiar with some of the actual and projected problems in scientific research and technological advancement; (b) gain experience in the analysis of ethical problems, and specifically, in the type of problems which they are likely to encounter during their professional careers, and (c) apply normative principles to these cases in order to be in a position to recommend actions and solutions.

In accomplishing this purpose, a subsidiary goal will be reached. The casebook should have the effect of drawing the attention of students and others to the variety, gravity, and pervasiveness of ethical problems in science. By "variety" here I mean simply that there are different sorts of problems that arise in the different sciences. Even within a science, there are different types of problems, and certainly there are problems which overlap the conventional boundaries which are normally used to mark off the individual sciences (e.g. subject selection is a problem for both psychology and medical research.)

By "gravity" I mean to point out that although some of the problems represented by these cases are more serious than others, the majority of cases raise issues that vitally affect the life and certainly the welfare of at least one individual if not that of large numbers of persons.

"Pervasiveness" means that there are ethical difficulties arising in all of the sciences. Furthermore, many of these problems have potential effects—for good or for ill—that will affect society and the world as a whole (e.g. pollution, nuclear weapons).

A third and hopefully immediate goal of the casebook is to begin in a more systematic manner discussion of these ethical problems in the hopes that continued research and discussion will lead to generally accepted solutions.

USE

The casebook is divided into three sections, each dealing with a different type of problem.

1. **Research.** This section deals with the ethical issues involved in scientific research itself and faced by the individual researcher or a small research team. These cases are concerned with both the ends and the means and procedures of the research in question.

2. **Scientist-Employer.** This section considers some of the ethical issues that can arise in research with respect to the relation between the individual scientist or research team and their employers, whether commercial, academic, or governmental.

3. **Science-Public.** This section raises some of the larger issues in scientific research and technological advancement, especially those which affect society as a whole and which must be decided by society as a whole with the guidance of scientists.

IX

CASES AND CASE ANALYSIS

The cases are described in as neutral terms as possible. *There is no overt implication of impropriety intended and none should be presumed from a case's appearance in the casebook or from the way it is described.* The whole point of the casebook is to decide—in so far as possible—on the ethical issues involved, not to presuppose that an indictment has already been given. Obviously, if a case did not raise some potentially interesting questions, it would not have been included; but this should not be taken as a list of moral wrongs committed by science.

A case analysis should include all of the following items.

1. **Description of the Case.** When the case is one given in this casebook, the description given should be considered complete. When a supplementary case is being analyzed, be certain to include in your description all of the potentially relevant factors. This is very important, since in any discussion of the cases, it must be presumed that the description is complete. Otherwise there will be needless disagreement resulting from one party knowing or presuming factors or circumstances which are not in the description.

2. **Bibliography of the Literature on This or Similar Cases.** It is important to take into consideration in any case analysis what has been said about the problem by others. This will also provide sources for further research on the problem.

3. **Analysis of Ethically Relevant Factors.** As has been already pointed out in the previous chapter on ethics, not all of the circumstances and factors in a case are ethically relevant. Thus, before any analysis can begin, it is necessary to isolate those factors which must be taken into consideration in arriving at both a clear discussion of the ethical issues at stake and in offering

an ethical judgment. Here we should note if there is any disagreement between our list of ethically relevant factors and that of precedent analyses. Unless there is agreement on this point, the whole of the rest of the analysis will remain in question.

4. **Relevant Normative Principles.** It will be useful at this point to list the normative principles which may be appealed to in justification of your judgment. In addition to the principles which you accept, include the principles which might be appealed to by someone who accepts a different normative system than yours. This is quite important since in many cases, a judgment would be justified by the normative principles of several different systems. For example, some ethical judgments may be justified by utilitarian principles, intuitionist principles, natural law principles, and axiomatic principles. In other cases, where different principles would justify different judgments, it is important to isolate this as the focus of disagreement.

5. **Your Considered Ethical Judgment.** Here the case analyst should offer his judgment on what ought to be done by the individual researcher, the review committee, his colleagues, the employer, or the public, as the case demands. The ethical judgment should be justified by appeal to the relevant normative principles and the ethically relevant factors in the case.

6. **Discussion.** This part includes a discussion of any divergent views found in the literature, along with an attempt to isolate the point of disagreement. In addition, any other possible judgments which can be justified by the same normative principles appealed to in part five should be included. Also, the judgments that would be made by those committed to different principles than yours should be noted. Lastly, this discussion should consider what relevant factors could or must be changed in order to change the moral judgment offered in part five. For example, what safeguards, modifications of experimental procedure, qualifications of researcher, or circumstances could be initiated or changed that would lead to a different moral judgment. Thus, if an experiment, for instance, is found to be morally unacceptable, what could be changed to make it acceptable?

7. **Abstract.** At the end of an analysis, it is very useful to include an abstract of the case, discussion, and recommendation so that others can easily and quickly identify the analysis.

The point of view taken in case analysis should be that of individual investigator or member of ethics review committee, or interested citizen (where applicable) confronted with the particular case. Thus, discussion of the case should include relevant scientific information (with a summary understandable by a nonscientist), ethical considerations, discussion of any disagreements, and your personal recommendation based on all of the available evidence, both ethical and scientific.

The following is an example of case analysis.

1. **Description of Case.** "In a series of experiments designed to discover the effects of a student's feelings of success or of failure at a particular task, the experimenter artificially induced feelings of success and failure in different groups of subjects. In the failure experiment, a subject was asked to learn a rather complex motor task and the experimenter expressed surprise at how slowly the subject learned, compared his performance unfavorably with that of other students, and expressed sympathy with him for his clumsiness. The net result was to induce in the subject a feeling of inferiority and of self-derogation. By the end of the experimental session, some subjects were depressed, brooding, and angry, and had lost a measure of self-esteem." (*Privacy and Behavioral Research.* Washington, D.C., U.S. Government, Executive Office of the President, Office of Science and Technology, February, 1967, p. 12)

2. **Bibliography.** 23, 24, 25, 26, 27, 28, 35, 36, 40, 44, 52, 93, 107, 124, 146, 154, 309, 331, 343, 371, 385. (Numbers refer to text bibliography. Give full bibliographic reference to any source not included in annotated or supplementary text bibliographies.)

3. **Ethically Relevant Factors.** The primary question here is the justifiability of lowering or destroying a person's confidence and self-esteem. Also, it is not clear from the description how much the subject was told about the experiment. Thus, the question of informed consent arises. The most important factors are these.

a. Goal of the experiment
b. Necessity of lowering self-esteem
c. Severity of the failure feelings
d. Presence or absence of informed consent

In *Privacy and Behavorial Research,* only factor 'c' was considered and it was treated only from the point of view of rehabilitation. This, of course, is probably the central factor with reference to which the other relevant factors must be discussed. If it is possible at the end of the experiment to completely explain the experiment to the subject, to inform him that his performance was in fact quite acceptable, that it was necessary to downgrade his performance for experimental purposes, and, in general, to restore the subject's confidence and self-esteem, then the other factors will not be as important as if this were not possible.

But even if it is not possible to completely restore the subject's confidence, the experiment may be justified if the goal of the experiment is sufficiently important, and if the needed data can be obtained in no other way.

Whether rehabilitation is or is not possible, informed consent should be obtained. Here, as in many other cases in psychology, there can arise a significant disagreement as to how informed the subject must be in order that his consent be valid. If it is not possible for experimental reasons to completely inform the subject, then careful consideration must be given to the goal of the experiment and the necessity of this particular experimental technique if the procedure is to be even plausible from a moral standpoint.

If rehabilitation is not possible, the question of informed consent becomes paramount.

4. **Normative Principles.** The relevant normative principles in adjudicating this case are some or all of the following.

a. (Teleological) One ought to act so as to maximize utility to the greatest number of people. Here the utility value might be knowledge or the general welfare or benefit to be derived from the experimental data.

b. (deontological) One ought not to cause needless harm or suffering to others—including psychological harm.

c. (axiomatic) One ought not to do anything to decrease another's freedom.

5. **Ethical Judgment.** It is my considered opinion that the described experiment would be immoral unless two conditions could be fulfilled: (a.) the subject could be informed completely about the nature of the experiment, apprised of the possible psychological harm and risk, and (b.) the subject could be completely rehabilitated. The former condition might invalidate the experiment.

All of the listed normative principles serve to justify this judgment. Since it is not clear in this case that the derived data will be of paramount importance or will be of great practical benefit; and since it is not clear that the information cannot be gathered in any other way, I do not see a mandate from the utilitarian principle to perform the experiment. This especially so in view of the known severity of the psychological harm.

If, however, the experiment were to be justified, the utilitarian principle (a) would be the only applicable one.

On the deontological principle (b), the experiment is unjustified since its possible benefits would not be a sufficiently relevant consideration to allow violation of this principle. And, there is no doubt but that this experiment would be such a violation. Here we might mention that temporary psychological pain (during the experiment) is not nearly as important as possible long-range harm. For, no one denies the importance of confidence and self-esteem to a person's ability to adjust to his environment, his interpersonal relations, and to his general well-being.

The axiomatic principle (c) could not serve to justify the experiment since psychological harm (lowering of self-esteem) is just as much an impediment to personal freedom as is political or economic unfreedom. It is, perhaps, much more devastating to freedom since there is no external cause that the person can blame and since this impediment is notoriously difficult to remove.

6. **Discussion.** The discussion of this case in *Privacy and Behavioral Research* (*op. cit.*, p. 12) argues for the acceptability of such an experiment on two grounds: (a) "The subject can

normally recover his usual level of self-esteem . . ." and (b) ". . . the body of research of which this example is a part has led to a substantial modification of education policy in America." As we mentioned in Section 3 above, if rehabilitation can be easily and completely effected, then the other factors (e.g. consent) become less important. In addition, this discussion argues for the importance of the information received, which, again, is an important consideration and which could easily lead to a modification of our judgment. In fact, our judgment rested on a consideration of precisely these two points.

It is important to notice however, that in *Privacy and Behavioral Research,* the discussant emphasizes, ". . . it is the responsibility of the experimenter to make sure that this recovery occurs."

In this example, as in many others, the final judgment will hinge on the relevant factors rather than the principles involved. In other words, if we agreed that rehabilitation is possible and the results are of considerable importance, we would all probably agree that the experiment is acceptable. Lack of agreement on these two points will lead to a lack of agreement concerning the acceptability of the experiment itself.

7. **Abstract.** In a series of experiments designed to discover the effects of a students feelings of success or of failure at a particular task, the experimenter artificially induced feelings of success and failure in different groups of students. The two most relevant factors in this case are the possibility of complete rehabilitation and the value of the information gained. Experiment was judged acceptable only if rehabilitation is possible and only if the information gained outweighed the temporary harm to the subjects.

X

ETHICAL CASES IN SCIENTIFIC RESEARCH
(Research Questions)

These cases represent problems in the research itself and must usually be decided by the individual scientist or review committee. They have primarily to do with the goal of the research, but more importantly, with research methods and procedures. The cases in this section are primarily from psychology and medical research.

1. "In a study designed to discover the causes of personality qualities in children it was necessary to secure measures of the children's personalities. One device that has been widely used is the so-called sociometric measure which assesses certain personality characteristics of a child on the basis of judgments about him by his classmates. A set of statements about the child's behavior was prepared. Examples are: 'He usually suggests a good idea for a new game.' 'He always gets mad when we don't do what he wants.' 'He can read better than anyone else in the class.' The children were instructed to fill in the name of the child best described by each of these statements. By tabulating the answers given by all children in a class, it was possible to find out the peer judgment about various qualities of personality." (*Privacy and Behavioral Research*, op. cit. p. 11.)

2. "In a study designed to discover the relationship between level of anxiety and the need to be with someone, the investigator induced an anxiety state by deceiving his subjects. Without deception, he could not have obtained the levels of anxiety required to demonstrate this relationship." (*Privacy and Behavioral Research*, op. cit., p. 12.)

3. "In a study to discover the degree to which persons could be persuaded to inflict severe pain on others, subjects were led

to believe that they were administering electric shocks of considerable magnitude to other subjects. Many subjects were persuaded to increase the level of shock to points where the apparent subjects (who actually did not receive a shock) writhed in simulated pain." (*Privacy and Behavioral Research, op. cit.,* pp. 12-13.)

4. "In a study to discover how well a family can survive an extended stay in a fallout shelter, the investigator recorded all conversations during the interval, without the family's prior knowledge or consent." (*Privacy and Behavioral Research, op. cit.,* p. 13.)

5. In the course of research on concept formation, a new experimental task was developed which appeared to have promise in studying cognitive deficits in schizophrenia. The research was not likely to have any therapeutic value. The subjects were patients in a VA hospital. (Professor Robert Haygood, Dept. of Psychology, Kansas State University, Manhattan, personal communication.)

6. Certain well-known universities require introductory psychology students to participate in experiments. Often the student can choose which experiments to participate in. Also, frequently, experimental participation is not required, but class credit is given for it. Often the experiments are directly connected with class material. (Haygood, *loc. cit.*)

7. "We recently ran an experiment involving probabilistic feedback. Such feedback creates a problem which cannot be "solved" in the ordinary sense, and hence leads to a situation which may constitute a 'failure experience' for the subject. Furthermore, because of the possibility of subjects passing the information on, it was deemed advisable that the subject *not* be informed that the problem did not have a perfect solution. We did, of course, tell the subject that the problem was 'extremely difficult,' and that hardly anyone ever did find the complete answer. While the risk of failing to solve, or learn, is always present in human learning experiments, it may be another matter entirely to subject a subject deliberately to such an experience, especially without debriefing.

"Note that the subjects usually cannot tell that the problem has no solution—most subjects believe that the solution is merely a great deal more complex than they have been able to unravel.

"Our purpose was not to look at failure experiences; these were simply a necessary by-product of the type of problem used." (Haygood, *loc. cit.*)

8. In an experiment to study task performance under conditions of great anxiety, students were placed at tables, all in view of all others. Anxiety was induced in the following way. An apparatus was electrically connected to the skin of the subjects. The subjects were told it would measure homosexual response to pictures flashed on a screen in front of the subjects. If strong homosexual responses were received a red light would go on on the subjects table. In fact, the experimenter controlled the "queer" lights. After this portion of the experiment was completed, the subjects were given a standard task performance test.

9. In order to test the possible chemical factors involved in schizophrenia, a drug was administered to normal subjects. Their behavior then closely paralleled that of schizophrenics. When the drug wore off, various after effects were noticed, and these after effects were of varying severity. The experiment was performed several times with different groups of subjects in an effort to eliminate both the after effects and as a control.

10. A sexual adjustment and maturity test was given a group of college freshmen. The results of the test were kept in departmental files for one semester, then discarded.

Sub-case: The results were given to the subjects themselves.

Sub-case: The results were given to the counselling center.

11. In a study of the effects of packing and temperature on confined groups of people, a bomb shelter was simulated. Subjects were completely informed of the goal of the experiment and consent was obtained. However, subjects were not told that the entire bomb shelter period was recorded on video tape. Later the tape was provided to another experimenter studying the dominance-dominated factor in interpersonal relations.

12. In tests on the effects of a psychodelic drug, the experimenter also took the drug. This was necessitated, according to

the experimenter, in order to communicate with the subjects since memory reports of the effects on the part of the subjects was found unreliable.

13. In a study of peer group (or superior group) conformity, one subject is placed with a group of stooges. The stooges give erroneous answers to relatively simple questions (e.g. estimating the size or weight of objects). Then the subject is asked the same questions, unaware that the others deliberately erred.

14. A researcher is given a grant from an advertising association to study the techniques and effects of subliminal conditioning.

15. "This study was directed toward determining the period of infectivity of infectious hepatitis. Artificial induction of hepatitis was carried out in an institution for mentally defective children in which a mild form of hepatitis was endemic. The parents gave consent for the intramuscular injection or oral administration of the virus, but nothing is said regarding what was told them concerning the appreciable hazards involved." (Beecher, Henry K., M.D.: Ethics and clinical research. *New England Journal of Medicine*, June 16, 1966)*

16. "Live cancer cells were injected into 22 human subjects as part of a study of immunity to cancer. According to a recent review, the subjects (hospitalized patients) were "merely told they would be receiving 'some cells'— . . . the word cancer entirely omitted . . ." (Beecher, *op. cit.*)

17. In a study of the effectiveness of a new multipurpose vaccine against childhood diseases, permission was sought and granted to use wards of a county court in a children's home as subjects.

---

*Reprinted from *New England Journal of Medicine* by permission of Dr. Henry Beecher, M.D. and the editors of the *Journal*. Copyrighted June, 1966.

18. "It is known that rheumatic fever can usually be prevented by adequate treatment of streptococcal respiratory infections by the parenteral administration of penicillin. Nevertheless, definitive treatment was withheld, and placebos were given to a group of 109 men in service, while benzathine penicillin G was given to others.

"The therapy that each patient received was determined automatically by his military serial number arranged so that more men received penicillin than received placebo. In the small group of patients studied, 2 cases of acute rheumatic fever and 1 of acute nephritis developed in the control patients, whereas these complications did not occur among those who received the benzathine penicillin G." (Beecher, *op. cit.*)

19. "The sulfonamides were for many years the only antibacterial drugs effective in shortening the duration of acute streptococcal phryngitis and in reducing its suppurative complications. The investigators in this study undertook to determine if the occurrence of the serious nonsuppurative complications, rheumatic fever and acute glomerulo-nephritis would be reduced by this treatment. This study was made despite the general experience that certain antibiotics, including penicillin, will prevent the development of rheumatic fever.

"The subjects were a large group of hospital patients; a control group of approximately the same size, also with exudative Group A streptococcus, was included. The latter group received only nonspecific therapy (no sulfadiazine). The total group denied the effective penicillin comprised over 500 men.

"Rheumatic fever was diagnosed in 5.4 per cent of those treated with sulfadiazine. In the control group rheumatic fever developed in 4.2 per cent.

"In reference to this study a medical officer stated in writing that the subjects were not informed, did not consent and were not aware that they had been involved in an experiment, and yet admittedly 25 acquired rheumatic fever. According to this same medical officer *more than 70* who had had known definitive treatment withheld were on the wards with rheumatic fever when he was there (sic.)." (Beecher, *op. cit.*)

20. "During bronchoscopy a special needle was inserted through a bronchus into the left atrium of the heart. This was done in an unspecified number of subjects, both with cardiac disease and with normal hearts.

"The technic was a new approach whose hazards were at the beginning quite unknown. The subjects with normal hearts were used, not for their possible benefit but for that of patients in general." (Beecher, *op. cit.*)

21. "The percutaneous method of catheterization of the left side of the heart has, it is reported, led to 8 deaths (1.09 per cent death rate) and other serious accidents in 732 cases. There was, therefore, need for another method, the transbronchial approach, which was carried out in the present study in more than 500 cases, with no deaths.

"Granted that a delicate problem arises regarding how much should be discussed with the patients involved in the use of a new method, nevertheless where the method is employed in a given patient for *his* benefit, the ethical problems are far less than when this potentially extremely dangerous method is used 'in 15 patients with normal hearts, undergoing bronchoscopy for other reasons.' Nothing was said about the granting of permission, which was certainly indicated in the 15 normal subjects used." (Beecher, *op. cit.*)

22. In order to test the toxicity and the effects of a new radiation treatment, subjects who were classed as 'incurably ill' or 'terminal' were used. The reasons given for this choice of subject were two: (1) although the new treatment was thought to be ineffective in patients at this stage of development of the disease, any improvement in these patients would be good grounds to conclude efficacy of treatment for patients in the initial stages of the disease, and (2) since treatment involved grave risks, the use of 'terminal' patients is justified.

23. In experimentation to develop a cure for a new strain of malaria, a double-blind technique was used, giving some patients the new drug, while others received placebos.

24. "To study the sequence of ventricular contraction in

human bundle-branch block, simultaneous catheterization of both ventricles was performed in 22 subjects; catheterization of the right side of the heart was carried out in the usual manner; the left side was catheterized transbronchially. Extrasystoles were produced by tapping on the epicardium in subjects with normal myocardium while they were undergoing thoracotomy. Simultaneous pressures were measured in both ventricles through needle puncture in this group.

"The purpose of this study was to gain increased insight into the physiology involved." (Beecher, *op. cit.*)

25. In tests to discover the effects of a new drug to treat schizophrenia and alcoholism, no prior animal experiments were conducted. The reason given was that schizophrenia and alcoholism are specifically human ailments and animal experiments would provide no useful information.

26. It is current practice in the drug industry to have practicing physicians participate in the final clinical evaluation of new drugs. These physicians are not qualified experimenters and the use of their patients for this research usually does not permit adequate control or followup studies.

27. A well-known biological research institute, in conducting studies in human reproduction, purchases and surgically obtains live embryoes and fetuses.

28. A biological research group, for studies in human reproduction, purchases and surgically obtains ovaries and testicles from living donors who are adults and properly execute forms of consent.

29. In this experiment, the circulatory system of a cancer patient is connected with that of a kidney disorder patient. The reasoning here is that the kidneys of the cancer patient can purify the blood of the kidney patient, while the kidney patient may produce antibodies to counteract the tumors in the cancer patient. Such an interconnection of two organisms is called "parabiosis."

30. Electrodes are implanted in the brains of rats in order to test the arousal of characteristic states of aggression, sexual drive, hunger, and rest.

31. While working on certain aspects of paper and pulp chemistry, an investigator discovers a flavor enhancer which can easily be extracted from the bark of certain trees. The investigator also finds that the extract is especially suited for use in the manufacture of cigarettes.

Sub-case: From previous research, the investigator is convinced of the deleterious effects of smoking on health.

XI

ETHICAL CASES IN SCIENTIFIC RESEARCH
(Employer-Employee Questions)

Cases in this section are designed to illustrate two types of problem: (1) problems which result from or relate to the employer-employee relationship, whether the employer is a university, a corporation, a private research group, or the government, and (2) ethical problems which go beyond the confines of the individual researcher and his experimentation and require the joint consideration of both the researcher and his employer. There are many ethical issues which are not represented here because they are germane to any employer-employee relationship and are, thus, not restricted to this relationship where the employee is a scientist. For example, consulting for competing firms, industrial espionage, employment practices, etc., are not here represented not because they are not important ethical questions, but because they do not affect scientists alone.

1. In a local hospital, a research group was involved in heart research involving catheterization of the heart and multiple x-rays of the heart. Yet no heart-lung machine was available.

2. At a university, a psychologist conducts an experiment on perception in human subjects. Part of the procedure includes the administration of various drugs. The psychologist does not have a medical degree, and although a physician is available for consultation, he is not present during experimental sessions.

3. At a state university, the director of general research is the only person to review experiments, except in some cases, where they are reviewed by the head of the experimenter's department.

4. At a large chemical company, a researcher is given the choice of either participating in the company's Biological Chemical Warfare research or terminating his employment.

5. A nuclear physicist comes to a university with a govern-

ment grant to build a large electron accelerator. Upon his arrival, his colleagues pressure him to purchase immediately equipment from a company in which they all hold stock.

6. A university researcher makes a consulting agreement with a company working on a process practically identical to that being developed by one of his university colleagues.

7. A university offers a position to an industrial researcher. In the course of interviews with various members of the department, he accidentally divulges information useful to a current university research program. Later, the department as a whole vetoes his appointment.

8. Recently a large university announced an academic conference on nuclear physics and invited leading physicists to attend. However, all of the speakers and the funds for the conference were supplied by a manufacturer of equipment used in nuclear research.

9. "Starting out in his new job as production supervisor on a specialty chemical unit, Mason notices a high pressure reactor that is used occasionally for special orders. The reactor is not walled off from the rest of the process area, so he recommends that a concrete or sandbag barrier be erected around the vessel to protect the operators.

"Mason submits his recommendation to the plant manager, but no action is taken. When he finally inquires about the matter, he is told that the reactor has been run for five years without any trouble and such a barricade would be a needless expense. Mason insists, however, that in his best engineering judgment such a reactor is a definite hazard.

"After hearing Mason's arguments, the plant manager authorizes construction of a thin plywood wall around the unit—which Mason believes is merely an attempt to lull fears rather than to provide adequate protection. But having made considerable fuss already, he decides not to risk antagonizing the plant manager by pressing the issue further." (How useful are our ethical codes. *Chemical Engineering*, Sept. 2, 1963; with permission, courtesy of *Chemical Engineering*).*

---

*See Footnote page 88.

XII

ETHICAL CASES IN SCIENTIFIC RESEARCH
(Public Questions)

This section requires more of an introduction than did the other two sections because here the cases are much more complex and the responsibility for deciding them is divided. In the first section of the casebook (X), the ethical issues affected a relatively small number of persons and the decision rested primarily with the individual researcher—or at most with an ethical review committee. In the second section (XI), although the decisions are not the sole responsibility of the researcher, a small number of persons were involved and the effects of the proposed actions were not particularly grave or pervasive.

This, however, is not the case in this section. Here we want to discuss problems which will or are affecting society at large and which cannot be effectively decided either by the individual researcher or a small number of his colleagues.

The scientist, nevertheless, must take a leading role in the resolution of these cases for two important reasons: (1) he is the only one with the required information on the possible effects of certain decisions, and, (2) he at present holds an unrivaled position of social prestige which makes his voice louder, if not more correct, than that of others in society.

In all of these cases, the individual scientist must decide for himself, irrespective of what society decides, whether or not he will personally contribute to certain efforts or concur with certain positions. An example here would be the decision of some scientists not to contribute to Biological Chemical Warfare, even though this effort is apparently generally accepted.

Thus, in case analyses, part five (ethical judgment) should include what decision is recommended for society at large by you as a scientist but also it should include what you as an individual scientist should do when the general public takes a position at variance with your considered judgment. Notice also that because of the complexity of these cases, no definitive judgment may be possible. In this case, analysis may be limited to a discussion of the consequences of each alternative and of the ethical issues involved in each alternative.

In these cases, although the consequences of almost all of the possible alternatives are grave and quite pervasive, these effects are still in the future (with some exceptions such as organ transplants). Thus, your discussion should include the following two points. (1) What further information is needed to facilitate a resolution of the ethical problem, and (2) How should this problem be solved (decision mechanism) and in particular, by whom —if it is possible to single out one group of particularly well-placed individuals. For example, should the American Medical Association, the American Bar Association, the proposed government commission on medical ethics, or perhaps all three jointly, decide the question of when death occurs and attempt to lay down new guidelines which would be medically, legally, and morally acceptable?

1. So-called "mind-control" by chemical means has been around for a long time in the form of alcohol, stimulants, depressants, and psychedelics. However, none of these leave the reasoning powers intact; reasoning, effectiveness, and adaptability are greatly affected. However, suppose a chemical agent has been discovered which keeps people in a constant state of euphoria while not affecting their ability to reason and act. What are the issues involved in putting such a chemical in municipal water systems just as we presently flouridate water.

2. Subliminal conditioning is another way of affecting behavior without the knowledge of those whose behavior is being affected. Thus, suppose it is proposed by the National Institute of Health that the massive problem of emotional disturbance, aberrant behavior, and mental illness could be handled on a wide scale by use of subliminal conditioning via television, for example.

3. The implantation of electrodes in the brain has already

had some limited success with manic depressives. Thus, it is proposed that part of the legally sanctioned punishment of criminals be the implantation of electrodes which will effectively channel their behavior in socially acceptable ways. Secondly, with a black-out mechanism, the problem of escapees would vanish since any attempted flight could be stopped by radio-electronically blacking out the fugitive.

4. Cloning is the reproduction of identical offspring by culturing a body cell (as opposed to a germ-cell such as spermatazoa). Each cell contains all of the genetic information to completely reconstruct an identical organism. So far it has been possible to successfully clone carrots. Supposing cloning is technically feasible for humans, should I be allowed to clone myself? How many times, and this is an important question since cloning requires only one cell from the parent to produce a complete, identical offspring?

5. Eugenisis is the attempt to genetically improve offspring by careful control of genetic factors. Thus, should we not take ova from great women by artificial inovulation and spermatazoa from great men, effect conception and then reimplant the embryo in especially selected women? Perhaps in the near future, this implantation will become unnecessary with the advent of artificial placenta in which every biological condition can be carefully controlled throughout the entire gestation.

6. The technical ability to control the sex of offspring appears near. Thus, ought we not begin limiting population by means of drastically decreasing the number of women? Conversely, the population could be, if the need arose, dramatically increased in two generations merely by greatly increasing the number of females.

7. Brain centers which control the libido have been identified. We already possess mild aphrodisiacs and sexual depressants. Thus, it is proposed that birth control be achieved by chemically depressing the sex drive.

8. Although the question of organ transplantation in humans has been around for some time, it has recently been catapulted into the forefront by recent success and has captured the imagina-

tion of the popular press. Thus, rather than set this problem up in case form, we will restrict ourselves here to listing some of the legal and moral questions which remain to be answered.

 (a) Is it morally right for a normal healthy person to consent to the mutilation and impairment of his own body even when the motive is laudable (Kidney transplants, for example)?

 (b) What about the law of maim, dating from Medieval times which forbade the mutilation of one's body which rendered the person unfit for military service? In recent times, this law has been interpreted to allow surgical operations in the interest of one's own health; but how about the health of another?

 (c) How are we to decide the legal issue of consent when the donor is a minor?

 (d) Are organs commodities such that they may be bought and sold at will as other chattel?

 (e) In most states, no irrevocable gift of an organ can be made when the organ is to be taken from a person's cadaver. Next of kin can overrule a person's explicit wish to be a donor. Furthermore permission may not even be sought of a next of kin until after the prospective donor's death.

 (f) Because of the critical time factor, the question of when a person is dead has become of paramount importance.

 (g) "A particularly interesting case occurred in Britain in 1963, in Newcastle. A man was butted in the course of a fight and fell backwards on his head, causing severe cerebral hemorrhage. Fourteen hours after he was admitted to the hospital he ceased breathing and was placed on a respirator. Twenty-four hours later, with his wife's consent, a kidney was removed to be given to another patient. The respirator was disconnected and breathing and circulation ceased. The coroner's court considered whether the removal of the kidney had contributed to his death. On the surgeon's evidence that death was inevitable and that the man had only been placed on the respirator so that the kidney could be removed, the coroner committed the dead man's assailant on a charge of manslaughter." (G. R. Taylor, *The Biological Timebomb*, p. 74-75) In a similar case in Dallas, Texas in 1968, however, the district attorney publicly stated that it would be very difficult to prosecute the case since it was not clear whether the man was killed by the assailant or by the medical team removing his heart for a transplant in another patient.

 (h) Not only is the buying and selling of organs a distinct possibility, but blackmail or coercion in the procurement of organs may be expected.

(i) Can animal organs be used in transplants?

(j) Since possible recipients (over 40,000 maimed in automobile accidents and over 1,000,000 cardiovascular deaths in the United States yearly) so grossly outnumber possible donors, how do we decide not only who will be donor, but more importantly who will be recipient?

(k) What are the issues involved in ovarian or testicular transplantation? How about cosmetic transplantation, of ears, noses, or even breasts?

(l) What of the possibility of grafting human hands on primates in order to get low-level domestic slaves with opposed thumbs?

(m) What about creating chimeras for specialized purposes. For example, athletes with two hearts and three lungs, skilled technicians with three arms and hands?

9. Since 1963 researchers have been progressively more successful in keeping a decapitated animal brain alive by connecting it to the circulatory system of a normal animal. In response to the question of the feasibility of doing this with a human brain, Dr. Robert J. White answered: "There is no question that this is within the capability of laboratories today." (Taylor, *op. cit.*, p. 121) In fact, human brains pose less problems than animal brains since we have fully developed heart-lung apparatus for humans. When this technique becomes commonplace in the not too distant future, will physicians be obligated to maintain people alive in this manner?

10. According to one biologist, it is theoretically possible for a DNA extract from one ovum to be used as the fertilizing agent for another ovum. Thus, a woman could fertilize herself. A technique called, by the way, "auto-adultery" by Professor Jean Rostand. "The logical extension of this proposition is the complete elimination of men and the creation of a race of Amazons." "As the British physiologist Professor A. S. Parkes has observed: 'Women are beginning to have the scarcity value previously held by men. Biologically . . . there are something like a million tons of unnecessary male biomass in this country alone!' " (Taylor, *op. cit.*, p. 170-171) Thus, it is proposed that the sexual balance be changed genetically to favor the reproduction of females.

11. At present it is not possible to isolate and suppress genes which produce most defects. And, "such a defect can rather rapidly spread through a population, as is shown by the case of the way in which the disease known as Huntington's chorea was introduced into the North American continent. . . . In the seventeenth century, six people with this condition arrived in America. When a survey was made in 1916, 962 cases could be identified, including those no longer alive, and the way in which the gene had spread across the country from east to west, with steps of one or more generations on the way, could be traced. None of these 962 people need have suffered if the original half dozen could have been persuaded not to procreate." (Taylor, *op. cit.*, p. 174) It has already been proposed in the past (e.g. Chief Justice Holmes) and is presently being proposed that hereditary defects could be prevented and those genes removed from the gene pool by enforced sterilization of defectives.

12. Because of their inherent complexity, the following three problems could not be set forth in case form. However, the moral questions involved, as well as the political and economic problems, are too important to be omitted.

Biochemical agents have been periodically used in warfare for over 4,000 years. Throughout the history of their use they have been considered especially abhorrent weapons. Today, throughout the world, large financial resources are made available for the development and procurement of biochemical weapons. In fact, they have been used more widely today than since World War I (e.g. Vietnam, Yemen, and domestic riot control.) Biochemical warfare (and its development for that use) raises at least the following issues.

a) Is it morally justifiable to use BCW against large civilian populations? Under what circumstances?

b) Is there an ethically important distinction between lethal and nonlethal agents.

c) What about the use of defoliants and herbicides against crops and food supplies?

d) What are the issues involved in using biological agents against military opponents in combat situations?

e) Is there a viable distinction between development of offensive and defensive BC measures?

f) Should there be a rapid and effective effort to sharply curtail BCW research and procurement?

g) Is the risk entailed in research and development offset by military advantages of BCW?

h) There are, of course, other matters worthy of discussion on this topic (e.g. the argument that our national security requires BCW research) but I think they will come out clearly in discussion of items (a) to (g)

13. In many respects discussion of nuclear weapons will parallel that of BCW. However, there may be some important differences. If so, these should be brought out in discussion of the following points.

a) Is it morally justifiable to use nuclear weapons against civilian populations? How about the use of nuclear weapons against military targets even when the indirect result (fallout) will severely affect civilians?

b) Under what conditions would it be permissible to employ nuclear weapons, either tactical or strategic?

c) What are the issues involved in using the threat of nuclear weapons in the achievement of policy goals.

d) Is there any obligation on nuclear nations not to disseminate nuclear technology (with military applications) to non-nuclear nations?

e) Are the dangers inherent in the testing and development of nuclear weapons offset by their military or political advantages?

14. The last of the "large issues" I will set forth for discussion is the question of environmental pollution.

a) Irrespective of the political and economic reasons for controlling pollution, are there any moral reasons for doing so?

b) Do the manufacturers of nonbiodegradable substances such as plastic and certain detergents have any special responsibilities?

c) Do I, as an individual pollutant have any special responsibilities?

d) Our present cemetery practices are an increasingly severe source of land pollution. Ought we to change our methods of body disposal even over the religious objections of some?

e) How ought this whole problem be handled. In other words, are we under any obligation to settle this problem now?

15. Suppose a chemical company devised an addictive chemical that had only one property; it addicted a person to whatever it was in. Could that company then sell that chemical to, say, a peanut butter manufacturer? Since we do not now prohibit the sale of certain addictives, e.g. alcohol, cigarettes, why should we oppose the sale of other addictives?

XIII

CONCLUSION

The cases in the previous three sections were chosen because each raises a slightly different ethical issue and can provide a concrete foundation upon which discussion of the issue can be based. Hopefully, students and practicing scientists who read this casebook will contribute additional cases to a revised edition—especially when their new cases raise issues overlooked by my choice of cases.

The casebook, like the rest of these materials, should not be seen as a finished product, but rather as the beginning upon which others are encouraged to add.

As a word of caution well worth repeating, let me reiterate once again that these cases should not be taken as a bill of particulars in an indictment of science and scientists. Scientists, as I have said again and again, are no less moral than others. In fact, in view of the immense complexity of the moral problems they face, they are often clearly more concerned with these issues than others.

In Chapters X and XI, the ethical problems set forth for discussion raise current issues and represent the kind of moral questions which scientists must answer daily. In Chapter XIII, however, we have a short lag time in most of the cases to discuss and hopefully solve the moral issues before they are upon us. This, I think, is the central reason why we must begin this discussion now. Every month we procrastinate—in some cases, every day—some alternatives become no longer available. I think we have learned from the development of nuclear energy the undesirability of deciding the ethical issues ex post facto.

Thus, if this book is of any aid in the solution of current problems and a goad toward the discussion of future problems, then its purpose will be deemed well accomplished.

Part Three

Bibliographies

ANNOTATED BIBLIOGRAPHY

1. Abernathy, J. M.: Some ethical implications of professional planning. *American Society of Civil Engineers Journal of Professional Practice, 90 (No. 3896):*23-29, May, 1964.

   The author contends that the movement to establish "planning" as a profession unto itself carries ominous ethical implications. It is the responsibility, he argues, of the engineering profession to return planning to its proper place as an integral part of the total engineering operation.

2. Abernathy, James M.: Some ethical implications of professional planning. *American Society of Civil Engineers Journal of Professional Practice, 91(No. 2):*86-89, September, 1965.

   This closure is made to discussions of the author's article of the above title. The original article appeared in the May, 1964, *Journal;* the discussants' (Grava, Beer, Woolhiser, and Falkson) articles appeared in the January, 1965, *Journal.*

3. Adamson, Arthur: Letters—the scientist and the dominant danger. *Science, 133(No. 3460):*1271-1272, April 21, 1961.

   In this reply to Charles P. Snow's "Moral Un-Neutrality of Science," (*Science Digest,* March, 1961) the author is apparently saying that (a) scientists are no more ethical and ought not to be more concerned with ethical questions than anybody else, and (b) scientists should clearly and absolutely face "the dominant danger to the world today: Soviet power and aggressive intent."

4. Addinall, R. L.: Cowardly Patient. *Science, 153:*694, August 12, 1966.

   Commenting on P.J. Burnham's "consent form," (*Science,* April 22, 1966) the author concludes that if such forms were ever taken seriously, medical progress could be stopped.

5. Albert, M. L.: Vietnam: the doctor's dilemma. *Nation, 206:*823-824, June 24, 1968.

   The author's case, basically, is that physicians, because they have selected a vocation dedicated to the maintenance and preservation of life, have many reasons for, and would be practicing "preventive medicine" by becoming conscientious objectors.

6. Albrecht, William A.: "Man and his habitat: wastebasket of the earth." *Bulletin of the Atomic Scientist, XVII(No. 8):*335-340, October, 1961.

   Man is biologically contaminating himself and all the other

populations which support. Albrecht implies, but does not deal with, some questions of the ethical propriety of this mass "suicide."

7. Alderman, Frank E.: Ethics and municipal engineers in private practice. *American Society of Civil Engineers Journal of Professional Activities, 92(No. 1):*67-68, May, 1966.

The author discusses F. Pandullo's article of the above title which appeared in the September, 1965, *Journal.*

8. Alderman, M.: Medical experiments on humans; new guidelines. *New Republic, 155:*11, December 3, 1966.

The author discusses the natures of therapeutic and biomedically advancing experimentation in these reviews and criticizes the guidelines set up by the Public Health Service.

9. Alexander, R. L.: How to publish and perish. *Civil Engineering, 37:* 77-78, June, 1967.

The author advocates engineers participating in political engineering decisions and rejects the view that civil engineers should concern themselves solely with the mechanics of the city and leave to others the job of planning it.

10. Alexander, Shana: They decide who lives, who dies." *Life,* p. 102-125, November 9, 1962.

This article is a report of the shortage of kidney dialysis machines and the decisions made and problems confronted by "The Admissions and Policies Committee of the Seattle Artificial Kidney Center at Swedish Hospital," or what the author terms, "Seattle's Life or Death Committee."

11. Allison, H. C.: AAAS defines social role of scientists. *Bulletin of the Atomic Scientist, 16(No. 9):*302, September, 1960.

The author presents a report of a statement by the American Association for the Advancement of Science relating to the social responsibilities of scientists.

12. American Medical Association: Ethical guidelines for clinical investigation. *Today's Health, 45:*70, April, 1967.

Four major guidlines, conforming to and expressing fundamental principles of both the World Medical Association's *Declaration of Helsinki* and the American Medical Association's *Principles of Medical Ethics,* for clinical investigation are listed.

13. Anonymous Management Consultant: Gathering competitive intelligence, *Chemical Engineering, 73:*143-148, April 25, 1966.

This report on methods of gathering competitive information raises many ethical questions for professional engineers as well as other scientists.

14. Auger, P.: Scientist looks at popularization. *Unesco Courier, 15:*14-17, June, 1962.

The author discusses certain trends relative to the popularization of science.

15. Authors and editors: Corruption in a certain segment of American
medicine, as presented in *The Healers, Publishers Weekly, 191*:40,
February 13, 1967.
A review of *The Healers* by Anonymous, M.D. (Putnam)

16. Bachrach, Arthur J.: The ethics of Tachistoscopy. *Bulletin of the
atomic scientist, XV (No. 5)*:212-215, May, 1959.
Bachrach gives a concise nontechnical explanation of subliminal projection, outlines the arguments in favor and in opposi-
tion to its use, and finally contends that the use of subliminal advertising is not in keeping with accepted ethical standards (APA).

17. Baker, Jeffrey: Letters—Science: philosophical problems. *Science,
151(No. 3713)*:935, February 25, 1966.
A letter in reply to B. Glass of December, 1965 which says,
"A science that welcomes group subjectivity is as dangerous as one
which fails to recognize the presence of subjectivity within it."

18. Baker, William O.: The moral un-neutrality of science—comments.
Science, 133(No. 3448):261-262, January 27, 1961.
Baker holds that since what the scientist may or may not do is
determined by "public morality," the scientist must make every
effort to gain public trust and understanding.

19. Barber, B.: Resistance by scientists to scientific discovery. *Science,
134*:596-602, September 1, 1961.
This article questions certain practices within the scientific
community.

20. Barber, B., and Walter, H. (Editors): *Sociology of Science*. Glencoe,
Illinois, Free Press, 1962.
In this textbook of readings, the editors divide the material into
the social nature of science and the scientific role, the reciprocal relations between science and society, Soviet scientists and the great
break, the social image and self-conceptions of the student, the
organization of scientific work and communication among scientists,
the social process of scientific discovery, and finally, the social
responsibilities of science.

21. Beals, Ralph L. Cross-cultural research and government policy. *Bulletin
of the Atomic Scientist, XXIII(No. 8)*:18-24, October, 1967.
This was concerned with the effect of government sponsorship
(e.g. Department of Defense) and government subversion (e.g.
CIA) in anthropological tests and studies.
The following are "indefensible."
1. Research classified as secret.
2. "Secrecy regarding sponsorship, source of funds, or the objectives of research."

3. The problems of consent and protection of individual not separated from activities of CIA.

4. No social scientists in review-of-project levels in the Foreign Affairs Research Council.

22. Becker, W. C.: More on ethics: protection for the employer. *Chemical Engineering Progress, 61*:33-35, April, 1965.

The author presents and defends the employer's point of view of ethics in relation to the subject of trade secrets and confidential information.

23. Beecher, Henry K.: Correspondence—human experimentation. *New England Journal of Medicine, 275(No. 14)*:791, October 6, 1966.

This is a rejoinder to replies by Silverman, Katz, Wessel, and Scott (*et. al.*) to the author's Ethics and clinical research. (*New England Journal of Medicine, 274(No. 25)*:1354.

24. Beecher, Henry K., M.D.: Documenting the Abuses. *Saturday Review,* 49:45-46, July 2, 1966.

This article is the same as that which appeared in the *New England Journal of Medicine* with the exception that twenty separate cases of violation of patient's rights from the medical literature are excerpted. According to *Saturday Review,* "Dr. Beecher's brief description of the experiments was for the most part uninformative to the laymen."

Dr. Beecher argues that the two main ethical components of experimentation in man are first, the informed consent of the patient and second, the presence of an intelligent, informed, conscientious, compassionate, responsible investigator.

He discusses, in detail, the difficulty of obtaining "informed consent" but holds that it is "absolutely essential to strive for it." At the very least, the subject must know and agree to be a participant in an experiment.

The intelligent investigator is necessary to insure that the experiment does not become ethically jeopardized by thoughtlessness and carelessness.

"An experiment is ethical or not at its inception;" he concludes, "it does not become ethical *post hoc*—ends do not justify means."

25. Beecher, Henry K.: Ethics and clinical research. *New England Journal of Medicine, 274(No.42)*:1354-1360, June 16, 1966.

The author, Dorr Professor of Research in Anesthesia at Harvard Medical School, limits this study to the category of "experimentation on a patient not for his benefit, but for that, at least in theory, of patients in general."

In this famous paper, Dr. Beecher deals with the urgency of ethical study; discusses the frequency of ethical violations and the problem of consent, and lists twenty-two case examples. He concludes that the two most important ethical components are an informed consent and an ethical investigator; that no ethical distinctions between ends and means exist in experimentation; and that data unethically obtained should not be published.

26. Beecher, Henry K.: Consent in clinical experimentation: myth and reality. *American Medical Association Journal, 195(No. 1)*:34-35, January 3, 1966.

The author labels as myth the notions that consent is freely available, that doctors would not make any request to a patient not for that patient's "good" and that the end justifies the means.

27. Beecher, Henry K.: *Experimentation in Man.* Springfield, Illinois, Thomas, 1958.

This book deals with human experimentation in terms of the subject; the investigator; subject and investigator relationships; justification for human trial; types of human experimentation; permissible vs. not permissible experimentation; propriety in publication; ethical and moral aspects; legal consideration, and codes in existence.

28. Beecher, Henry K.: Some guiding principles for clinical investigation. *American Medical Association Journal, 195(No.13)*:1135-1136, March 28, 1966.

The author offers guidelines for human experimentation in the areas of normal volunteers, self-experimentation, patient volunteers, patients requiring therapy and in experimentation on patients not for their direct good but for the welfare of patients in general.

29. Beer, Charles G.: Some ethical implications of professional planning. *American Society of Civil Engineers Journal of Professional Practice, 91(No. 1)*:59-60, January, 1965.

The author replies to J. M. Abernathy's article of the above title which was published in this *Journal,* May, 1964.

30. Bennet, J.: Letters—grantitis. *Science, 139*:235, March 22, 1963.

This letter is a reply to C. J. Flora's "Grantitis" in *Science,* December 7, 1962.

31. Bergen, Richard P.: Law and medicine—racial problems in medical practice. *American Medical Association Journal, 195(No. 12)*:299-301, March 21, 1966.

The author considers racial factors in hospital staff privileges, hospital patients, membership in professional associations, and private office practice.

32. Berding, A. H. Crucial decade, *U. S. Department of State Bulletin, 43*:671-676, October 31, 1960.

The author discusses certain problems concerning the role of science in public policy for the future.

33. Bergner, Lawrence: Letters—"Captain Levy and the army system." *Science, 157(No. 3784)*:140, July 14, 1967.

    In this reply to Elizabeth Langer's "Court Martial of Captain Levy: Medical Ethics vs. Military Law," (*Science*, June 9, 1967) Berger criticizes Langer for not discussing "the eternal question of accepting responsibility for the ultimate use that is made of one's research," and contends that the Army went "out of its way" to put Levy in jail because he raised just such a question.

34. Berkley, Carl: Ethical dilemmas in medical engineering, *American Journal of Medical Electronics, 5(No. 1)*:9-10, 1st quarter, 1966.

    The author lists some cases of ethics in medical engineering and suggests that a possible solution might be the formation of an Ethics Committee.

35. Bishop, Jerry E.: Ethics, risks of experimentation on human patients cause increasing concern in the medical profession, *Wall Street Journal*, p. 6, August 31, 1964.

    The author claims that much of the problem of ethics of experimentation is due to the fact that persons in experiments are not healthy volunteers but sick patients hoping to be treated. He discusses the Nuremburg Code and specific experimental problems and questions.

36. Bixler, Ray H.: Experiments on humans—the growing debate: "Ostracize them!"—a challenge to professional societies. *Saturday Review*, 49:47-48.

    The author details common examples of ethical violations and suggests that these could be eliminated by the ostracizing of the violators by professional colleagues and organizations.

37. Blaker, C. W.: Thanatopsis. *Christian Century*, 83:1503-1506, December 7, 1966.

    The author discusses ethical issues concerning "death" and the ethical environment within which such discussion must take place.

38. Bochard, Mrs.: Hearings to resume on health science commission—exhibit 1 (letter). *Congressional Record, 114:(No.46)*, March 20, 1968.

    The author, who states she "worked in a hospital surgical unit for several years," argues that there are many serious moral and ethical questions in medicine and supports a commission to investigate such.

39. Bockus, Frank M.: Letters to the science editor—medical ethics. *Saturday Review*, 49:51, September 3, 1966.

    Dr. Bockus states that what is at stake in the medical ethics question is "the place of values in scientific methodology." He also refers to the fact that an interdisciplinary conference (National Conference on Human Genetics and Bio-Chemistry) was held.

That conference called for continuing conversation on the philosophical and ethical problems implied in biomedical advances.

40. Bolinger, R. E.: Medical experimentation on humans. *Science, 152*: 448, April 27, 1966.

The author claims that for ethical reasons to be considered appropriately, human experimentation must be divided into two categories, "observational" and "manipulative." The observational category uses only treatment, whether diagnostic or therapeutic, that would be used in nonexperimental situations. Ethical questions of this category are well covered by established codes. Manipulative experimentation, however, is the application to a patient of a risk posing procedure which cannot conceivably benefit him. Ethical and legal aspects of manipulative experimentation have yet to be resolved.

41. Bolt, R. H.: Statesmanship in science. *Physics Today, 14*:30-32, March, 1961.

The author considers, among others, the aspect of the relationship between science and public policy.

42. Born, Max: Physics and politics. *Bulletin of the Atomic Scientist, XVI (No. 6)*:194-200, June, 1960.

Both the scientist and the military planner must place ethical limits on certain military actions (nuclear devices). The primary responsibility for each is that civilization continue to exist.

43. Brickman, W. W.: Science, liberal arts, and the national crisis. *School and Society, 90*:101, March 10, 1962.

The author considers the relationship of the academic community and public policy and discusses the role of science within that framework.

44. Britain's Medical Research Council to Parliament: Medical ethics. *Science 145*:1024-1025+, September 4, 1964.

Excerpts from the statement—"Responsibility in Investigation on Human Subjects," dealing with procedures contributing to the benefit of the individual, control subjects in investigations of treatment or prevention, procedures not of direct benefit to the individual, and professional discipline.

45. Brode, W. R.: Development of a science policy. *Science, 131*:9-15, January 1, 1960.

The author considers the relationship on a policy level, between science and the government.

46. Brode, W. R.: Growth of science and a national science program, *American Scientist, 50*:1-28, March, 1962.

The author emphasizes the necessity of a greater involvement and interaction within public policy by science.

47. Brode, W. R.: National and International Science, *U. S. Department of State Bulletin,* 42:735-739, May 9, 1960.

The author considers some of the relationships between public policy, national science, and international science.

48. Brode, W. R.: Role of science in foreign policy planning, *U. S. Department of State Bulletin,* 42:271-276, February 22, 1960.

The author considers the necessity for the increased use of science in matters of foreign policy planning.

49. Bronk, D. W.: Idea that science can solve everything is a false one. *U. S. News and World Report,* 48:75-76, February 22, 1960.

The author discusses certain popular misconceptions about science.

50. Bronowski, J.: Moral for an age of plenty. *Saturday Evening Post,* 233:24-25+, November 12, 1960.

In a nondetailed article, the author claims that science and scientists have a very effective code of morality in that, "we ought to behave in such a way that we can all find out what is true."

51. Bronston, William G., M.D.: The physician and Vietnam. *Bulletin of the atomic scientist,* XXII(No. 9):24, November, 1966.

Dr. Bronston states that it is the ethical responsibility of war physicians and the health profession to document the carnage and suffering of war. According to his view, apparently, doctors should serve the war effort but should not do so with "closed eyes and hearts."

52. Bross, Irwin D. J.: Experiment—or stagnate. *New York Times Magazine,* p. 4, July 23, 1967.

This is a reply to Goodman's article of July 4, 1967. He argues that Goodman overlooked two essential items relative to the question of experimentation: (a) a patient in a well designed clinical experiment often receives overall medical care that is far superior to routine practice, and (b) the alternative to experimentation is stagnation.

53. Bruce-Chwatt, L. J.: Letters—experimentation: rights and risks, *Science,* 155(No. 3770):1617-1618, March 31, 1967.

In this reply to Wolfensberger's "Ethical issues in Research with human subjects," (*Science,* p. 47, January 6, 1967), Bruce-Chwatt holds that the malaria research using consenting, informal, human patients (prisoners) was more ethical than "some trials done on hospital patients without their knowledge or consent."

54. Budd, John H.: Medical ethics—what is wrong with fee splitting? *American Medical Association Journal,* 195(No. 2):117-118, January 10, 1966.

The author discusses the "gray" cases of fee splitting and sets forth the decisions by the American Medical Association in various cases.

55. Bundy, McGeorge: Scientist and national policy. *Science, 139*:805-809, March 1, 1963.

This is an address made by the author December 27, 1962, in which he discussed the role of the scientific community in the administration of the government.

56. Burnham, P. J.: Medical experimentation in humans. *Science, 152*:448, April 22, 1966.

The author ridicules the "patient's rights" issue legally developed in the New York Regents vs. Southam-Mandel case, by presenting a satirical "consent" form.

57. Bush, V.: Other fellows ball park. *Science, 134*:1163, October 20, 1961.

This editorial discusses the role of scientists when entering the "other fellows 'ball park' " of affairs of government and politics.

58. Calder, R.: Science reporter speaks of babelology, *UNESCO Courier, 14*:46-47, July, 1961.

The author considers verbosity, incommunicability, verbage, and redundancy in the scientific community.

59. Campbell, T. L.: Reflections on research and the future of medicine. *Science, 153(No. 3734)*:442-449, July 22, 1966.

The author lists new possibilities of medical science in genetics in that, for all practical purposes, genes can be manufactured to certain specifications. This certainly seems to raise ethical questions but the report does not consider them.

60. Can scientists find the clue to peace? *Business Week*, 140-142+, September 15, 1962.

This article considers certain aspects of scientific involvement in public policy, including foreign affairs.

61. Carley, William M.: Patient consent to research: rules set. *Wall Street Journal*, p. 12, January 21, 1966.

The author reviews the legal basis of the Southern-Mandel case in which the two doctors (Southam and Mandel) were found guilty of fraud, deceit and unprofessional conduct in an experiment involving the use of live cancer cells in elderly patients without the patient's knowledge that the cells being injected were cancerous.

62. Charyk, J. V.: Scientist and his responsibility. *Aerospace Engineering, 20*:7, July 7, 1961.

The author considers the ethical and social responsibility to his society.

63. Chase, E. T.: Politics and technology. *Yale Review, 52*:321-329, March, 1963.

The author discusses the political influence of the scientific community as well as the political implications of scientific advances.

64. Chemical World 1959/1960: Science in government. *Chemical and Engineering News,* 38:63-64, January 4, 1960.

    The report discusses the political nature of certain aspects of science.

65. Christensen, N. A.: Function of ethical codes. *Petroleum Management,* 37:72-75, October, 1965.

    The two main reasons for "the painstaking studies of engineering ethics" are, first, that a professional code of ethics is a necessary base for a profession; and second, that technology has created different groups in engineering, each with its own unique problems. Two charts of ethical canons and situations are given.

66. Christensen, N. A.: Purpose of professional engineering ethics, *Mechanical Engineering,* 87:46-48, November, 1965.

    The author considers five professional engineering societies and lists codes and particular types of ethical problems peculiar to each.

67. Cisler, W. L.: Engineer in international affairs. *Mechanical Engineering,* 82:54-56, February, 1960.

    The author discusses the role of the professional engineer in matters of foreign policy.

68. Classen, H. George: Fact and purpose. *Bulletin of the Atomic Scientist,* XXIV(No. 3):36-38, March, 1968.

    The author says that science, "the search for truth," is not an ethical goal, but is a characteristic of man. He argues against ethics from evolution. He concludes that "Ethics (or purpose, or the self) is not subject to scientific definition." This avoids ethical values on science, but by implication would reject them.

69. Clyde, G. D.: Engineer and Politics. *Civil Engineering,* 30:45-46, August, 1960.

    The author discusses the role of the professional engineer in public planning and decision-making.

70. Code of ethics. *Naval Engineers Journal,* 77:45-47, February, 1965.

    This is the new code of ethics for engineers as approved by the Board of Directors of the National Society of Professional Engineers in July, 1964. It replaces the previously adopted, "Ethics for Engineers."

71. Code of ethics as amended to be voted on by ASCE membership. *Civil Engineering,* 31:42-44, June, 1961.

    The Code of Ethics of the American Society of Civil Engineers, proposed amendments, and the ASCE's "Guide to Professional Practice Under the Code of Ethics" are given.

72. Cohen, Edward: Criticism and the advancement of building engineering. *Civil Engineering,* 37:34-35, July, 1967.

    The author argues that not only is it permissable, but it is ethically obligatory that the professional engineer "seize" every opportunity for self, professional, and public criticism.

73. Commission on Health Science and Society: *Summary of Testimony.*
    This report summarizes the testimony of Dr. John Najarian, Dr. Adrian Kantrowitz, Dr. Arthur Kornberg, Dr. Joshua Ledecberg, Dr. Christian Bernard, Dr. Wangensteen, and Dr. Henry Beecher relative to the establishment of a Health Science Commission.

74. Committees of scholars support candidates; scientists joining, *Science,* *132*:1238, October 28, 1960.
    This report deals with the public support of certain candidates in the 1960 election by scientists.

75. Commoner, Barry: *Science and Survival.* New York, Viking Press, 1966.
    The author emphasizes many of the ethical problems stemming from recent scientific advances.

76. Conferences on science and world affairs, statements by participants; with editorial comment. *Science, 134*:971, 984-991, October 6, 1961.
    A report of the conferences on science and world affairs. An editorial opinion is also given.

77. Consent required for drug experiments. *Science News, 90*:172, September 10, 1966.
    This article reports the federal guideline that physicians obtain the written consent of patients for the use of investigational drugs.

78. Cowen, D. L.: Ethical drugs and medical ethics. *Nation 189*:479-482, December 26, 1959.
    The author deals here with the ethics involved in the pricing of prescription drugs and, in addition, the sources of information about new drugs to physicians.

79. Coswa VIII Statement. *Bulletin of the Atomic Scientist, 17(No. 11)*: 395-396, November, 1961.
    A statement of the conference on science and world affairs held at Stowe, Vermont.

80. Cranberg, Lawrence: An object of concern and study. *The Virginia Quarterly Review, 41(No. 4)*:653-656, Autumn, 1965.
    In the course of a review of Henry Margenau's *Ethics and Science,* the author contends that there are many serious ethical problems in science that Margenau simply did not deal with.

81. Cranberg, Lawrence: Ethical code for scientists. *Science, 141*:1242, September 27, 1963.
    Cranberg compares engineers and scientists in that the professional engineers have seen the need for, and adopted a code of ethics while scientists have done neither.
    The "thoughtful attention" of the scientist, scientist educator,

and scientific professional organization must now be given to the problem of a code of ethics, he concludes.

82. Cranberg, Lawrence: Ethical code for scientists? *Science, 142:*1257, December 6, 1963.

> In a rejoinder to W. E. Graham (*Science*, December 6, 1963) who advocated the rules of the Society for Social Responsibility in Science serve as a code of ethics, the author declares that the concern of SSRS with "a limited special range of ethical problems and its existence apart from the main body of professional scientific organizations only emphasize the disparities which exist between scientists and other occupational groups with respect to ethical education and regulation."

83. Cranberg, Lawrence: Ethical problems of scientists. *American Scientist, 53(No. 3):*303A-304A, September, 1965.

> The author lists ten examples of ethical problems in science and suggests that scientists ought to study the ethical-regulatory systems of other professions to be able, hopefully, to develop a professional ethic for science.

84. Cranberg, Lawrence: Ethical problems of scientists, *The Educational Record*, 282-294, Summer, 1965.

> The author discusses various types of ethical questions and problems and presents a case for the adoption, by scientists, of a professional code of ethics.

85. Cranberg, Lawrence: Ethical problems of scientists. *The Eleventh Edward G. Budd Lecture*, presented at the Franklin Institute, November 2, 1966.

> The author discusses some legal questions but primarily emphasizes "the problems of the ethics of scientists in relation to one another to their employers, and to society generally."

86. Cranberg, Lawrence: Science and technology—a time to respond to the public trust. An invited talk given to the Engineering Council for Professional Development. New Orleans, September 30, 1968.

> The main thrust of this talk is that "ethical sensitivity and moral perceptiveness" must be imparted by the colleges and universities to future scientists and technologists.

87. Cranberg, Lawrence: Science, ethics, and the law. *Zygon Journal of Religion and Science, II (No. 3):*262-271, September, 1967.

> The author analyzes the relationship between science and ethics and states that scientists may be lax for not having developed their own professional ethic.

88. Cranberg, Lawrence: The P. M. bomb. *Bulletin of the Atomic Scientist, XXIII (No. 12):* December, 1967.

> The author holds that the "power cum money" bomb—the explosion of resources into the sciences produced among scientists

an immediate sense of "social responsibility" in certain areas of public policy but that the "bomb" has many adverse consequences within the sciences which are not fully recognized by the public, or even many scientists themselves.

89. Crichton, J. M.: Heart transplants and the press. *New Republic, 158* (No. 21):28-34, May 25, 1968.

In the course of a review of *The Transplanted Heart* by Peter Hawthorne, Crichton considers the ethics of press involvement in medical operations.

90. Culliton, B. J.: Consent: it's the law. *Science News, 92*:88-89, July 22, 1967.

Culliton briefly traces a history of the "consent" requirements of the U. S. Food and Drug Administration and lists some problems stemming from the law and some possible solutions (including an insurance plan to cover patients and volunteers) to them.

91. Curran, W. J.: Privacy, birth control, and "uncommonly silly law." *New England Journal of Medicine, 273(No. 6)*:322-323, August 5, 1965.

The author discusses Griswold vs. the State of Connecticut (85SCt. 1678 [1965]) in which the Connecticut law barring the use of contraceptives was declared unconstitutional, on the reasoning of "the right of privacy."

92. Curran, W. J.: Problem of consent: kidney transplantation in minors. *New York University Law Review, 34*:891-898, 1959.

The author provides a commentary on three decisions to allow kidney transplants of minors of the same family and raises legal considerations on the questions of the age of the minors and the value of psychiatrist's testimony in general.

93. Curran, William: The law and human experimentation. *New England Journal of Medicine, 275(No. 6)*:323-325, August 11, 1966.

The author discusses some of the legal complications of human experimentation, especially the question of "informed consent."

94. Dafler, James R.: Letters. *Bulletin of the Atomic Scientist, XX(No. 10)*: 22, December, 1964.

In reply to John Haybittle's "Ethics for the Scientist," (*Bulletin of the Atomic Scientist,* May, 1964) Dafler says a scientist should not have "his conscience bother him excessively" for a decision made by a collective.

95. Dam, K. W.: Scientist and conflict of interest. *Bulletin of the Atomic Scientist, XXIII(No. 10)*:343; October, 1961.

The author discusses the relationship between conflict of interest statutes and certain practices within science.

96. Damon, V. G.: Fee splitting knife happy surgeons and mercenary

doctors. Excerpt from I learned about women from them. I. Taves (Editor). *LOOK, 26*:86-88+, June 19, 1962.

The author, a veteran M.D., denounces the unethical procedures and doctors listed in the title.

97. Danielson, Lee E.: *Characteristics of engineers and scientists.* Ann Arbor, Michigan, the University of Michigan, 1960.

This book is one of a series of reports attempting to identify and publicize managerial practices and policies that promote productivity and increase the satisfactions of engineers and scientists.

98. Dash, J. Gregory: Where responsibility lies. *Bulletin of the Atomic Scientist, XXIII(No. 1)*:35-37, January, 1967.

The scientist cannot "disclaim responsibility for the implications of one's scientific contributions." He argues that in World War II, the scientists got so caught up in the excitement of building the bomb, they forgot the *reason* they were building it: thus, it was used when it did not need to be. He says the same mistakes are happening today. Science is amoral; scientists provide the ethics (if any) and must bear and be assessed the responsibility for doing so.

99. Davis, A. S.: Ethics and the engineer, and what the law has to say about both of them. *Product Engineering, 36*:70-75, July 5, 1965.

This article is from a lecture delivered to the New York State Society of Professional Engineers. The author discusses the ethical implications of two predominant questions: "Can I take it with me?" (when accepting a new job); and "What is the worth of my services if I create something new?"

100. De Bakey, Michael E.: Medical research and the golden rule. *American Medical Association Journal, 203(No. 8)*:574-576, 1968.

De Bakey lists and explains certain guidelines for the physician, acting as experimenter and the *clinical* experimenter.

101. Dedner, S.: Why did Daedalus leave? *Science, 133*:2047-2052, June 30, 1961.

The author argues that underdeveloped countries are aiding the developed by exporting one of their most precious commodities, scientific talent.

102. Dedijer, S.: Research: The motor of progress. *Bulletin of the Atomic Scientist, XVIII(No. 6)*:4-7, June, 1962.

The author considers certain aspects of scientific research.

103. Dedijer, S.: Window shopping for a research policy. *Bulletin of the Atomic Scientist, XV(No. 11)*:367-371, November, 1959.

The author considers certain policy matters of scientific research.

104. De Leon, Benjamin: Is science morally sterile? *Bulletin of the Atomic Scientist, XXIV(No. 5)*:54; May, 1968.

De Leon argues that science, by espousing moral neutrality is actually unscientific because it fails to recognize that morality and ethics are a part of man's life and cannot be ignored. He says the moral neutrality comes from capitalistic emphasis on science.

105. De Solla Price, Derek J.: Ethics of scientific publication. *Science, 144(No. 3619)*: 655-657, 1964.

He discusses the availability of the literature, freedom to publish, the awarding of credit, citation, retrieval, invisible colleges, and scholarship.

106. De Solla Price, Derek J.: *Little Science, Big Science.* New York, Columbia University Press, 1963.

In this book, the author covers a prologue to a science of science, Galton revisited, invisible colleges and the affluent scientific commuter, and political strategy for big scientists.

107. Dickel, Herman A.: Medical ethics—the physician and the clinical psychologist. *American Medical Association Journal, 195(No. 5)*: 121-126, January 31, 1966.

In this comparison, the education and interrelationship between the physician and the clinical psychologist, the author discusses gaps in the realm of responsibility, the medical curriculum, the "newer discipline" of clinical psychology, differences in orientation and the interdependence and mutual respect of the two groups.

108. Dinsmore, R. P.: It's the individual's responsibility. *Chemical Engineering Progress, 61*:38, April, 1965.

The author argues that it is the professional responsibility of the engineer to be aware of ethical issues and situations and, also, to meet the professional standards in use.

109. Dobzhansky, T.: *Mankind Evolving.* Yale University Press, New Haven, 1962.

The author presents mankind as "a product of evolution and as an evolving whole" and deals with many ethical questions of science within this realm.

110. Doctors Dilemmas. *Scientific American, 218*:49-50, March, 1968.

This editorial deals with a symposium, "The Cost of Life," by the Royal Society of Medicine in England. According to the symposium's summary, doctors face two kinds of dilemmas: first, "the determination of priorities when limited resources make it impossible to provide all necessary medical care and, second, the decision as to when to discontinue artificial aids to survival." The symposium speakers considered different aspects of these dilemmas and some general guidelines were given.

111. Donnelly, James F.: Hearings to resume on health science commission
—Exhibit 1 (Letter). *Congressional Record, 114*:46, March 20,
1968.

> The author, president of Patients' Aid Society, Inc., says, in
> part, "I certainly don't think the decisions of human transplants
> should be left entirely to the doctors anymore than declaration of
> war should be left entirely to the generals."

112. Dowling, H. F.: Human experimentation in infectious diseases. *Journal
of the American Medical Association, 198*:997-999, November 28,
1966.

> "Tyrannical restraints on experimentation would certainly have
> precluded the invaluable contributions of Vesalius, Harvey, Jenner,
> Walter Reed . . . Rigid prescriptions about human experimentation
> would also have prevented the remarkable discoveries during the
> past century regarding the cause, transmission, and prevention of
> infectious diseases."

113. Dubos, R.: Scientist and the public. *Science, 133*:1207-1211, April 21,
1961.

> The author considers the relationship of the scientist, as
> scientist, and as a member and integral part of society.

114. Du Bridge, L. A.: Policy and the scientists. *Foreign Affairs, 41*:571-
578, April 11, 1963.

> The author considers the relationships between science and
> government.

115. Du Bridge, L. A.: Science and a better America. *Bulletin of the
Atomic Scientist, 16(No. 10)*:340, October, 1960.

> The author considers some of the ways in which science can
> interact more fully with the American society.

116. Du Bridge, Lee A.: The government role in science education. *Bulletin
of the Atomic Scientist, XXII(No. 5)*:16-20, May, 1966.

> Cited as an article dealing with the question of government
> support of scientific research and mentioning no ethical problems
> or consideration; by implication, they do not exist.

117. Earle, William: Ethics as technology. *Science, 147(No. 3654)*:140-141,
January 8, 1965.

> This is a review of Margenau's *Ethics and Science*, in which
> the author's thesis is "the book remains an elementary and con-
> fused effort."

118. Edel, A.: Science and the structure of ethics. In *International En-
cyclopedia of Unified Science. Foundations of the Unity of Science,
2(No. 3)*, Chicago, University of Chicago Press, 1961.

> The author divides this work into divisions of the nature and
> complexity of the problem; the theory of existential perspectives;

the role of science in conceptual and methodological analysis; and decision, freedom, and responsibility.

119. Eichenlaub, J. E.: Psychologic props: the truth about doctors' deceptions. *Science Digest, 51*:53-57, February, 1962.

The author discusses psychologic cures for physical diseases and suggests steps by which patients can help their doctors help them.

120. Eiduson, B. T.: Letters. *Science, 131*:552-555, August 26, 1960.

The author replies to Gerald Holton's "Modern Science and the Intellectual Tradition," in *Science*, April 22, 1960. Rejoinder by Holton.

121. Engineers speak out on ethics. *Chemical Engineering, 70*:177-184, December 9, 1963.

In this followup article to "How useful are our ethical codes," (September 2, 1963), the tabulation of the results of a poll conducted in that earlier article are given and interpreted.

122. Ethical problems in engineering. *Petroleum Management, 37*:76-77, October, 1965.

This article consists of two ethical problems and discussion of each. The problems and discussions are taken from *Ethical Problems in Engineering* (by Alger, Christiansen, Olmstead. New York, Wiley Co., 1966) which lists and discusses 127 ethical cases.

123. Ethics in business. *Chemical Engineering Progress, 61*:13-25, February, 1965.

This is a series of articles dealing with the question of ethics as related to the *business* of chemical engineering and not as related to the scientist.

124. Ethics of human experimentation, (editorial). *British Medical Journal, 2*:1-2, July 6, 1963.

This editorial considers some of the problems involved in "consent" in human experimentation and presents the first five points of the Nuremburg Code for Permissible Human Experiments.

125. Ethics of human experiments. *Time, 88*:42+, July 8, 1966.

A review of an article in *New England Journal of Medicine* by Dr. Henry K. Beecher in which Dr. Beecher expressed concern at experiments designed to benefit society but which may be harmful to the specific patient(s). *Time* listed several of Beecher's examples and offered a rebuttal for one.

126. Ethics of science evolved with man. *Science News, 89*:299, April 23, 1966.

This article briefly notes the evolutionary ethic of science in that it teaches man to adapt to his environment and to control his environment to meet all his own needs; then reviews *Science and Ethical Values* by Bentley Glass.

127. Etzioni, Amitai: When scientists testify. *Bulletin of the Atomic Scientist, 20*:23-26, October, 1964.

 The author holds that scientists, testifying before governmental committees are considered to be "disinterested" but, in reality, they oftentimes constitute a pressure group concealing sociopolitical interests by a practice of "sandbagging."

128. Executive Council of the Episcopal Church: Hearings to resume on health science commission—Exhibit 1, Resolution. *Congressional Record, 114(No. 46),* March 20, 1968.

 This resolution, adopted February 22, 1968, lists certain ethical problems stemming from new developments in medical technology and urges scientists to "exercise the greatest caution in human experimentation."

129. Experiments on man. *Newsweek, 69*:84, March 6, 1967.

 This news article discusses issues of legal consent in human experiments. It concludes that the final responsibility depends on the conscience of the investigator.

130. Experiments on man. *Scientific American, 215*:44, August, 1966.

 This is an editorial concerning Henry Beecher's article in *The New England Journal of Medicine.*

131. Fabro, J. A.: Death—what is at the heart of it. *The National Catholic Reporter,* 6, June 26, 1968.

 The author, an M.D., discusses the ethical questions relating to death and transplants.

132. Failure to communicate seen as science crisis. *Science News Letter, 80*:240, October 7, 1961.

 Irwin Hershey, publication director for the American Rocket Society said, this article reports, that American scientists are failing to communicate adequately with either themselves or the public.

133. Falk, Charles E.: Science and public policy activities in universities. *Bulletin of the Atomic Scientist, XXIV (No. 6)*:50, June, 1968.

 Enough work has not been done on "Science and Public Policy Activities in Universities," and we need more. Two examples are technology assessment and an "assessment of the role which science has played in the role of the social, economical, political, and cultural welfare . . ."

134. Farber, L. H.: Psychoanalysis and morality. *Commentary, 40*:69-74, November, 1965.

 The author's thesis is that psychoanalysis is, among other things, a moral science whose involvement with problems of good and evil is inescapable and essential. She argues that the psychoanalyst must take a moral position in the therapeutic context and should recognize that fact.

135. Federation of American Scientists: Classified research in the university. *Bulletin of the Atomic Scientist, XXIII (No. 8)*:45-46, October, 1967.

"The increasing dependence of American universities on the federal government could lend to the loss of this basic function of the university. (open inquiry, debate, and criticism, free and detached . . .) The dependence comes dangerously close when a university takes on classified military research." The guideline is that a university should accept only those projects which will be freely published and available.

136. Federation of American Scientists: The war and weapons in Vietnam. *Bulletin of the Atomic Scientist, XXIII(No. 5)*:59-60, May, 1967.

The authors are opposed to present roles in Vietnam and advocate the use of "restraint".

137. Finland, Maxwell: Ethics, consent, and controlled clinical trial. *American Medical Association Journal, 198(No. 6)*:637-638; 1966.

The author's basic point seems to be that no code or legislative body can competently decide on proper risks in experimentation as physicians are not always in agreement.

138. Flora, C. J.: Grantitis, *Science, 138*:1185-1186, December 7, 1962.

The author discusses the awarding of grants to scientists and raises questions about certain ethical abuses of such.

139. Fosberg, F. R.: Letters—"Code of ethics." *Science, 142*:916, November 15, 1963.

The author suggests, in reply to L. Cranberg (*Science, 141*: 1242) that scientists do have a workable code of ethics, "the principle of 'scientific honesty' and the complete realization that this is the very essence of science."

140. Fozzy, Paula: New scientific consultant policy. *Bulletin of the Atomic Scientist, XVIII(No. 5)*:43-44, May, 1962.

The author considers aspects of the new scientific consulting policy.

141. Fozzy, Paula: Scientists social responsibility. *Bulletin of the Atomic Scientists, XVIII(No. 3)*:45-46, March, 1962.

This news article lists the actions and recommendations of the "Committee on Science in the Promotion of Human Welfare" of the American Association for the Advancement of Science since 1955.

142. Franklin, R. K.: Questions of transplants. *New Republic, 158(No. 11)*:7, March 16, 1968.

After discussing some of the ethical aspects of *donating* and *receiving* hearts for transplantation, the author contends that ethical decisions made in the realm of transplants will form the basis for those to be made in the realms of genetics, human experimentation, and eugenics.

143. Freed, J. Arthur: *Some Ethical and Social Problems of Science and Technology: A Bibliography of the Literature from 1955.* U. S. Department of Commerce, Washington, D. C., 1964.

This bibliography is from 1955 to 1963, and is not annotated.

144. French and Japanese Scientists on War Research. *Bulletin of the Atomic Scientist, XXIV(No. 3)*:35, March, 1968.

Four hundred and thirty-three scientists signed a statement issued on February 5, 1967, that appealed to their "American colleagues" to refuse to work for war purposes. It said, in part, "Scientists have vast ethical and professional responsibilities in the modern world. If they are to work for the benefit of humanity and the integrity of scholarship, they should never willingly permit their discoveries to be exploited for destructive purposes."

145. Fruton, J. S.: Aims and values of the sciences. *Yale Review, 51*:197-210, Autumn, 1961.

Fruton's main point is that science, as a discipline, has inherited a cultural tradition of knowledge for its own sake from philosophy and that, even though it is largely overlooked because of many practical benefits, forms the basis of science. The discoveries of scientists are, then, community and not private property. Thus science as a discipline cannot be held responsible for uses of scientific discoveries, whether for "good" or "bad".

146. Furlong, W. B.: How doctors use patients as guinea pigs. *Good Housekeeping, 161*:79+, October, 1965.

The author lists certain ethical rights of patients and details many cases of ethical abuse. The author concludes by offering guidelines for human experimentation and ways for a patient to detect experimentation.

147. Galton, L.: Doctors debate fee splitting. *New York Times Magazine,* 19+, March 4, 1962.

The author examines ethical problems concerning fee splitting, the "current fuss about the nature of fee splitting in general," how it works and how and why it is as prevalent as it is.

148. Garceau, O.: Morals of medicine. *Annals of the American Academy of Political and Social Science, 363*:60-69, January, 1966.

The author discusses the nature of the medical ethic and recent developments of medical improvements, ethical advances, and new ethical questions.

149. Gerard, R. W.: Vivisection: ends and means. *American Institute of Biological Sciences* (now *Bio-Science*), *13(z)*:27-29, 1963.

The author discusses the "basic ethical questions of ends and means" as the root of many objections to animal experimentation.

150. Gingold, Kurt: Letters—'Captain Levy and the Army System'. *Science, 157(No. 3784)*:140, July 14, 1967.

In reply to Elizabeth Langer's "Court Martial of Captain Levy: Medical Ethics vs. Military Law" (*Science,* June 9, 1967), Gingold argues that a physician has no other rights than a citizen and in our society, each citizen may not decide for himself what is a crime.

151. Glass, Bentley: Scientists in politics. *Bulletin of the Atomic Scientist, XVIII(No. 5)*:2-7, May, 1962.

The author discusses the role of scientists in matters of public policy.

152. Glass, Bentley: The ethical basis of science. *Science, 150(No. 3701)*: 1254-1261, December 3, 1965.

Glass argues for the following four ethical bases of science: complete truthfulness; complete honesty; fearless defense of scientific inquiry and opinion; and a full communication of scientific findings through primary publication, synthesis, and instruction.

153. Goodman, P.: Human uses of science. *Commentary, 30*:461-472, December, 1960.

The author assumes the position that science is morally neutral —it may be used for human or inhuman purposes. He advocates certain "human" uses.

154. Goodman, W.: Doctors must experiment on humans; but, what are the patients' rights? *New York Times Magazine,* p. 12-13+, July 2, 1967.

The author refers to H. K. Beecher's "Ethics and Clinical Research," *New England Journal of Medicine.* He analyzes "informed" consent and the myriad of ethical questions revolving around this concept.

155. Gorovitz, Samuel: Ethics and the allocation of medical resources. *Medical Research Engineering, 5(No. 4)*:5-7, Fourth Quarter, 1966.

The author, professor of philosophy at Western Reserve University and Case Institute of Technology, discusses many questions of medical attention that have not "gained the attention they deserve." He considers three kinds of resource-allocation (1. personal resources; 2. space and equipment resources; 3. national resources) then looks at dialysis treatment (for degenerated kidneys) as an ethical "case in point." He concludes by differentiating between questions about medical ethics and questions of medical ethics and argues that both must be squarely faced.

156. Governing Council, Federation of American Scientists: Scientists speak out. *Bulletin of the Atomic Scientist, XXIV(No. 5)*: 50, May, 1968.

"Except in time of national emergency, the university should not be a part of the military establishment . . ." The article gives identification of some practices "which tend to subvert the traditional role of the university." The university should not have military contracts. Two examples are Project Camelot and covert support of university project's by CIA.

157. Graham, W. E.: Ethical code for scientists. *Science, 142*:1257, December 6, 1963.

In reply to L. Cranberg (*Science,* 141:1242, 1963) the author states that there is an organization, the Society for Social Responsibility in Science, which does take its ethical responsibility seriously. He lists the four rules of the SSRS and states they are, in effect, a code of ethics.

158. Grant revoked. *Bulletin of the Atomic Scientist, XVII(No. 7)*:299, September, 1961.

On June 21, 1961, the NSF revoked a $3800 grant awarded to Edward Yellin, a graduate student (M.E.) at the University of Illinois. The grant was revoked less than two weeks after a charge on the House floor by Rep. Gordon Scherer (R-Ohio) (HUAC) that NSF was giving fellowship funds to a communist. Yellin had been a member of the Communist party from 1948 to 1957. The University of Illinois was convinced however, that he signed the Broyles oath and the allegiance declaration in good faith.

159. Grava, Sigurd: Some ethical implications of professional planning. *American Society of Civil Engineers Journal of Professional Practice, 91(No. 1)*:60-63, January, 1965.

The author replies to and discusses J. M. Abernathy's article of the above title which was published in this Journal, May, 1964.

160. Gray, A. W.: When an engineer leaves a company what are the rights and obligations of employee and employer? *Machine Design, XXXV*:133-134, March 28, 1963.

The author discusses the engineer's talents, knowledge, and use of competitive information, with regard for the ethics and legal situations of the engineer and the employer.

161. Gray, H.: Obligations of engineers. *Plant Engineering, XVII*:155, February, 1963.

The author discusses certain legal, ethical and social responsibilities of the professional engineer.

162. Gray, J.: Science, man, and society. *UNESCO Courier, XIV*:30-33+, July, 1961.

This article is an abridgement of an address by the author in which he discusses the relationship between the scientist and the society.

163. Green, Harold P.: AEC information control regulations. *Bulletin of the Atomic Scientist, XXIV(No. 5)*:41-43, May, 1968.

The author discusses and criticizes some new (December, 1967) security regulations used by the AEC. "Regrettably, the AEC has not faced up to the fundamental policy questions which are raised by the proposed regulations: (1) "whether security controls should be placed upon dissemination of information wholly privately developed," and (2) "whether private R & D in nuclear weapons and explosives and gas centrifuge technology should be suppressed." There are two fundamental truths: (1) hoarding of "secrets" won't deter other nations, and (2) the rate of technological progress in any area, and therefore, the national strength is 'f' (total number of individuals who bring their talents to bear in that area). See also #164 (below).

164. Green, Harold P.: The AEC proposals—a threat to scientific freedom. *Bulletin of the Atomic Scientist, XXIII(No. 8)*:15-17, October, 1967.

Green argues that the regulations proposed by the AEC on May 2, 1967 are "flagrantly interfering" with free scientific inquiry. The proposals were later modified and Green reviews the modifications in *Bulletin of the Atomic Scientist,* May, 1968.

165. Green, Harold P.: The new technological era: A view from the law. *Bulletin of the Atomic Scientist, XXIII(No. 11)*:12-18, November, 1967.

The author discusses certain technological advances and then considers some undesirable consequences of them.

166. Greenberg, Daniel S.: It's time for science to act its political age. *Bulletin of the Atomic Scientist, XXIII(No. 8)*:36-37, October, 1967.

The author looks at the government sponsorship of science as something of a necessity: he dismisses "military management of research" and presumably other areas of government interference out of hand. Science must now "compete effectively" for its share of government money.

167. Greenberg, Daniel S.: Science and foreign affairs: new effort underway to enlarge role of scientists in policy planning. *Science, 138*: 122-124, October 12, 1968.

The author discusses certain decisions made to increase the governmental participation of the scientific establishment.

168. Grund, C. B.: Ethics of the total environmental system. *Heating, piping, and air-conditioning,* 38:113-116, December, 1966.

In this rather detailed report the author stresses the consulting engineer's need for ethics on providing his client with full and objective information and fair recommendations.

169. Guard against pressures. *Science News Letter,* 79:p 46, January 21, 1961.

   A report of a speech by Dr. Barry Commoner in which Dr. Commoner warned that scientists must guard against a breakdown of their integrity by social and political forces.

170. Guinea pigs and people. (editorial). *Christian Century,* 79:975-976, August 15, 1962.

   This editorial raises questions dealing with the "human element" of drug testing. The main point of this piece, is that physicians unequivocally do not have the right to use uninformed and nonconsenting patients for experimentation with drugs.

171. Hailsham, L. R.: Imperatives of international cooperation. *Bulletin of the Atomic Scientist, XVIII(No. 12)*:18, December, 1962.

   The author discusses the necessity for joint work among scientists of different countries.

172. Hamblen, John W.: Preservation of privacy in testing. *Science, 151 (No. 3715)*:1174, March 11, 1966.

   The author suggests that the use of a computer can efficiently keep test subjects anonymous in certain kinds of psychological testing.

173. Harris, Morgan: *Cell Culture and Somatic Variation.* Holt, Rinehart, and Winston, Inc., New York, 1964.

   This is an extremely technical book, dealing with the field of somatic cell heredity. Ethical problems of genetic advances are raised.

174. Hartley, H.: Science and Government. *Chemistry and Industry,* 1478-1480, September 16, 1961.

   The author discusses the role of the scientific community in planning public policy.

175. Haskins, C. P.: Technology, science, and American foreign policy. *Foreign Affairs, 40*:224-243, January, 1962.

   The author argues for the participation in and use of the scientific community in foreign policy decisions.

176. Haybittle, J. L.: Standards of conduct. *Science, 136*:917-919, June 8, 1962.

   The author replies to the letters of Henry Kaplan and Alexander Whittenburg in *Science,* March 16, 1962. It is not enough, he says, to condemn Russian scientists who work on and develop atomic weapons. It must be realized that the individual scientist, of *any* country, must bear the responsibility for his work.

   ". . . Every individual scientist should critically examine what he is engaged upon in the light of such faith as he possesses and satisfy himself that the two are not incompatible.

177. Haybittle, John: Ethics for the scientist. *Bulletin of the Atomic Scientist, XX(No. 5)*:23-24, May, 1964.

Haybittle contends that ethics can be properly applied only to the scientist and not to science. He states that "where the possible uses of the end-product of any scientific work are known, then those scientists doing the work share a part of the responsibility for those uses whether they be good or bad." He further states that science itself cannot help formulate a choice of ends, but that religion ("using that term in its widest sense") can.

178. Headings, Lois: Book notes and reviews—a white mouse for the mines: moralities for technopolis II. *Business Horizons*, p. 101-118, Summer, 1967.

The author reviews and lists many new books on ethics as well as certain ethical theories. She deals briefly with the particular question of the ethics of the scientist.

179. Heart of the matter, (editorial). *The Tulsa Tribune,* reprinted in *Congressional Record, 114(No. 25)*: February 20, 1968.

This editorial commends Senator Walter F. Mondale for his bill to create a commission on the ethical and social implications of health research. Though the questions raised, the editorial states, are as old as *Brave New World,* they are now much more relevant.

180. Heitler, W.: Ethics of the scientific age. *Bulletin of the Atomic Scientist, XX(No. 8)*:21-23, October, 1964.

Heitler lists the technological, biological, and chemical advances which make a new (i.e. nontraditional religion) ethics necessary. He labels certain "biological applications which rest on partial knowledge of life" as irresponsible and holds that the root of that irresponsibility is that "the part is taken for the whole."

In the absence of a new scientific ethics, Heitler says, it is essential to "recognize that the main body of present scientific activity concerns only one aspect of nature—the material—and cannot be regarded as providing a full grasp of reality."

Finally, research is needed for a new conscience-directed scientific ethics having as one main point, "a respect for life."

181. Hershey, Nathan: Letters to the science editor—medical ethics. *Saturday Review, 49*:51-52, September 3, 1966.

The author feels that an objective criteria for deciding what to tell a patient in an experiment can be developed. He advocates finding out what surgeons tell surgical patients and using that as a precedent for what physicians should tell their patients.

182. Hesburgh, Theodore M.: Science and technology in modern perspective. *Vital Speeches, 28*:631-634, August 1, 1962.

In this address, the author considers the change in the nature of science due to scientific and technological advances.

183. Hesburgh, Theodore M.: Science as amoral; need scientists be amoral too? *Saturday Review, 46*:55-56, March 2, 1963.

The author argues that science itself is morally neutral, but that the practicing scientist must make certain ethical decisions.

184. Hesburgh, Theodore M.: The moral un-neutrality of science—comments. *Science, 133(No. 3448)*:259-261, January 27, 1961.

In this "comment" to C. P. Snow's "The Moral Unneutrality of Science," the author states that "science is by its nature neutral. It can be used for good or evil," but its use is directed by the scientist. Thus, the scientist must bear the ethical responsibility for his work.

185. High noon in Vermont. *Newsweek, 58*:92, September 18, 1961.

This article covers the conference on science and world affairs at Stowe, Vermont.

186. Hill, A. V.: *Ethical Dilemma of Science and Other Writings.* New York, Rockefeller Institute Press, 1960.

The author, in this book, discusses the ethical dilemma of science, trailing one's coat, about people, refugees, science in two world wars and science in the commonwealth.

187. Hill, Austin Bradford: Medical ethics and controlled trials. *British Medical Journal, (No. 5337)*:1043-1049, 1963.

In discussing controlled clinical trials, Hill argues that the statistician, in helping to devise such trials must be aware of ethical problems and must assume responsibility with the doctor.

188. Hirsch, Walter: Knowledge for what? *Bulletin of the Atomic Scientist, XXI(No. 5)*:28-31, May, 1965.

The classical view of the "scientific ethos" (universalism, communism, search for truth, and organized skepticism) is not enough, Hirsch argues, in this day of the "scientist-manager" and "scientist-politician." The "new ethos" of science must answer the title question in that scientists must be responsible for the effects of scientific progress on human welfare.

189. Hoaglund, Hudson: Some reflections on science and society. *Bulletin of the Atomic Scientist, XV(No. 7)*:284-287, September, 1959.

The ethics of science (objective, unfettered search for truth) may, by extension, be accepted by the nonscientific community.

190. Holman, Edwin J.: Osteopathy and the law. *American Medical Association Journal, 195(No. 10)*:283-284, March 7, 1966.

Under the subheading, "Ethical Policy," the author sets forth the official position of the AMA regarding professional relationships between medical doctors, hospitals, and osteopaths.

191. Holton, Gerald: Book reviews—science and ethical values. *Science,*
     *151(No. 3716):*1375-1376, March 18, 1966.
     A review of *Science and Ethical Values* by Bentley Glass.
192. Holton, Gerald: False images of scientists. *Saturday Evening Post,*
     *232:*18-19+, January 9, 1960.
     The author presents certain misconceptions about science and
     the individual scientist.
193. Holton, Gerald: Modern science and the intellectual tradition. *Science,*
     *131:*1187-1193, April 22, 1960.
     The author presents the intellectual ethic of modern science.
194. Honey, J. C.: Federalist paper for the 1960's. *Saturday Review, 43:*
     43-44, July 2, 1960.
     This article, in part, deals with the involvement of science in
     governmental public policy.
195. Horowitz, Irving Louis: The life and death of Project Camelot. *Trans-*
     *Action, III:*3-7, 44-47, 1965.
     The author outlines the beginnings of Project Camelot, the
     reasons for its cancellation, the public uproar it caused and, finally,
     discusses the ethics of the policy research that was undertaken in
     Project Camelot.
196. House Committee on Government Operations (Editors): *Special In-*
     quiry on Invasion of Privacy. Washington, U. S. Government
     Printing Office, 1966.
     This volume is the hearings held before a subcommittee of the
     House Committee on Government Operations and contains testi-
     mony of all witnesses before same.
197. House Subcommittee on International Organizations and Movements:
     Report No. 4 on Winning the Cold War: The U. S. Ideological
     Offensive. Washington, U. S. Government Printing Office, 1966.
     This report deals with the relationships between the Behavioral
     Sciences and the national security.
198. Houssay, B.: Plea for science. *Americas, 11:*1482, December, 1959.
     The author argues for a greater development of science and
     greater freedom from restrictions.
199. How useful are our ethical codes. *Chemical Engineering, 70:*87-90,
     September 2, 1963.
     This editorial presents a brief discussion of engineering ethics
     and then offers ten cases and asks the reader to respond for
     tabulation. See #121 above.
200. Hubble, Douglas: Medical science, society, and human values. *British*
     Medical Journal, February 19, 1966.
     The author discusses human values, the doctor-patient relation-
     ship; the doctor as societal agent; the doctor as administrator; and
     the doctor as investigator.

201. Hughes, Thomas L.: Scholars and foreign policy: varieties of research experience. *Department of State Bulletin,* 53:747-758, 1965.

    The author discusses the role of research in foreign affairs and some ethical questions entailed therein.

202. Hutton, James H.: Letters—the rights and wrongs of fee-splitting. *American Medical Association Journal,* 195(No. 11):187-188, March 14, 1966.

    In reply to John Budd's "What Is Wrong With Fee-Splitting," the author comments that a doctor should be able to spend his money as he sees fit (including payment to another doctor) just as a lawyer can; and that too much time and energy is spent on this question.

203. Hyman, W. A.: Medical experimentation on humans. *Science, 152 (No. 3724):*865, May 13, 1966.

    The issue of the Southam-Mandel case was only if the experimenters had the right to inject patients with live cancer cells without the patients' knowledge; the legal ruling was that they did not.

204. Imshenetsky, A. A.: Modern microbiology and the biological warfare menace. *Bulletin of the Atomic Scientist,* XVI(No. 6):241-242, June, 1960.

    Imshenetsky, director of microbiology for the Academy of Sciences of the USSR, holds that scientists and especially microbiologists throughout the world "should combat the preparations for biological warfare."

205. Industry's ethics could be better. *Space Aeronautics,* 39:11+, March, 1963.

    This "readers round table" panel discussed ethical problems in the aerospace industry. One big factor, some panelists said, was the lack of ethics in the government.

206. In public affairs has science been overemphasized to the detriment of engineering? *Product Engineering,* 31:24-25, March 14, 1960.

    This editorial distinguished between the involvement of "science" and "engineering" government and claims engineering has been slighted to the expense of science.

207. International comments—medical science, society, and human values. *American Medical Association Journal,* 195(No. 12):1081, March 21, 1966.

    This article is a review of an article of the same title by Dr. Douglas Hubble in the British Medical Journal (February 19, 1966). Dr. Hubble states that if medicine is to be humanistic it must acknowledge the individuality and uniqueness of each patient. Although a doctor's first duty is to his patient, Dr. Hubble also adds, the doctor, as citizen, may have certain duties that transcend that relationship.

208. Interpretation of the ASCE code of ethics. *Civil Engineering, 30*:25-27, January, 1960.

The official interpretation of the American Society of Civil Engineers "Code of Ethics" is given by the Board of Direction of the ASCE.

209. Is martyrdom ethical. *Science News Letter, 87*:214; April 3, 1965.

This is a news story of a speech to a conference on the problems and conplexities of clinical research by Dr. Henry K. Beecher of the Harvard Medical School. Dr. Beecher emphasized that "what seem to be breaches of ethical conduct in experimentation are by no means rare, but are almost, one fears, universal." Dr. Beecher cited eighteen examples, all but one being anonymous. He was challenged for this use of anonymity and nondocumentation by other physicians and this challenge was the stimulus for his later article in the *New England Medical Journal*.

210. Isler, Charlotte: The Bronx Junior High School sex quiz. *Saturday Review, XLIX(No. 6)*:64-65, February 5, 1966.

The author objects to the use of the Minnesota Multiphasic Personality Inventory on junior high school students without parental consent. In addition, she raises other ethical issues relative to personality testing of students.

211. John, E. Roy: The brain and how it changes. *Bulletin of the Atomic Scientist, XXI(No. 9)*:12-14, November, 1965.

Progress in brain research is being made which, if used in education, could have utility for normal individuals and those having inadequate brain function. However, John argues, the lack of a firm value system in education (rationality, humanitarianism, democracy) would hinder the applicability of any scientific research.

212. Johnston, Edgar G. (Editor): *Preserving Human Values in an Age of Technology*. Detroit, Wayne State University Press, 1961.

This book is comprised of a series of Leo M. Franklin lectures given at Wayne State University by Edward U. Condon, Henry Steele Commager, Francis Biddle, Louis L. Mann, and Edgar G. Johnston.

213. Jolly, J. B.: Needed: stronger engineering groups. *Chemical Engineering, 71*:112-116, August 3, 1964.

One of the reasons that engineers are not as professional as doctors, the author claims, in this argument for professionalism, is that they have allowed nonprofessionals to work alongside. This of course, destroyed a strong base for effective ethical codes.

214. Jordan, P., and Keating, K. B.: Scientist in politics: on top or on tap?

Summary of debate. *Bulletin of the Atomic Scientist, XVI(No. 1):* 28-29, January, 1960.

The authors summarize the debate as to whether scientists should be "on tap" to the government or "on top" of it.

215. Judicial Council, American Medical Association: Professional courtesy survey. *American Medical Association Journal, 195(No. 4):*299-301, January 24, 1966.

This is a report of a survey of the medical profession's attitude toward professional courtesy. The summary is, "there has been little change in recent years."

216. Kanfer, Frederick H.: Issues and ethics in behavior manipulation. *Psychology Reports, XVI(No. 1):*187-196, 1965.

The author discusses three features of the psychotherapeutic process which tend to raise ethical problems: the particular methods of control used; the domain of the behavior to be controlled, and the discrepancy between personal values and cultural metavalues.

217. Kapitza, P. L. Future of science. *Bulletin of the Atomic Scientist, XVIII(No. 4):*37, April, 1962.

The author gives his consideration to future developments and policies within science.

218. Kaplan, Henry S.: Standards of ethical conduct. *Science, 135:*997-998, March 16, 1962.

The author submits a quotation of his letter to Professor V. Zhdanov of the Soviet Union. The quotation lists his reason for not attending the Eighth International Cancer Congress in Moscow as outrage and indignation at the resumption of atmospheric bomb testing by the USSR.

219. Katz, Jay: Correspondence—human experimentation. *New England Journal of Medicine, 275(No. 14):*790, October 6, 1966.

The author, Associate Professor of Law and Associate Clinical Professor of Psychiatry at Yale, commends Henry Beecher's "Ethics and Clinical Research" and adds two comments. First, that the nonpublication of "'improperly obtained'" data (as suggested by Beecher) would preclude review and appraisal of the conflicting values in experimentation. Second, that medical schools should have "intensive seminars" on ethical questions.

220. Kennedy, John F.: President sends message to conference on science and world affairs. *U. S. Department of State Bulletin, 45:*533, October 2, 1961.

This article is the text of President Kennedy's message emphasizing international cooperation in science.

221. Kenyon, Richard L.: A lesson from the lemmings. *Chemical and Engineering News, 5,* July 15, 1968.

This editorial suggests that technological advances may be ecologically harmful and that one of the great problems at present "is that of finding ways to choose directions of technological progress that will bring benefits without heavy social costs."

222. Kepes, Gyorgy: The research frontier—where is science taking us. *Saturday Review, 49*:66-67, March 5, 1966.

"We lack the depth of feeling and the range of sensibility needed to retain the riches that science and techniques have brought within our grasp." Thus, we need artistic values to place upon science to relieve our "environmental," "social," and "inner chaos."

223. Kidney transplants. *Nature, 217*:595, February 17, 1968.

This is a report of a private conference then forthcoming on kidney transplants. The article touches very briefly on the ethical and legal aspects on kidney removal after death without consent.

224. Killian, J. R., Jr.: Making science a vital force in foreign policy. *Science, 133*:24-25, January 6, 1961.

The author presents his case for increasing the scientific involvement in foreign affairs policy.

225. Killian, J. R., Jr.: Science and engineering: resources for peace. *Bulletin of the Atomic Scientist, XVIII(No. 3)*:2-5, March, 1962.

The author emphasizes the need for participation by science in government as a tool for achieving peace.

226. Kingdon, Frederick: Letters—Scientists indulged. *Science, 145*:873, August 28, 1964.

The author feels that it is not ethical to force the public to pay taxes to support certain projects which "offer nothing in return" to that public, save to allow certain scientists to indulge their whimsy.

227. Kistiakowsky, G. B.: National policy for science. *Chemical and Engineering News, 40*:120-122+, January 22, 1962.

The author emphasizes the role of science in public policy.

228. Kistiakowsky, G. B.: Science and foreign affairs. *Science, 131*:1019-1024, April 8, 1960.

This address, made January 29, 1960, stresses the need for the greater participation of science in foreign affairs.

229. Klausler, A. D.: Radiation and social ethics. *Christian Century, 80*:199-200, February 13, 1963.

The author discusses the ethical responsibilities of the practicing scientist.

230. Knock, Frances E.: Ethical problems of human experimentation. *American Geriatrics Society Journal, XIII(No. 6)*:515-519, 1965.

In the field of cancer chemotherapy, uses of sensitivity tests that may in some cases be harmful and other clinical practices violate sound chemical principles, the Nuremburg Code, and the Statement by the Medical Research Council. A more explicit and binding ethical code is needed.

231. Koop, C. E.: What I tell a dying child's parents. *Readers Digest, 92*: 141-145, February, 1968.

The author claims that doctors have more than their medical duty of doing their professional best. They also have a duty to provide "support and reassurance" to patients and relatives.

232. Kornberg, Arthur: Hearings to resume on health science commission—exhibit 1 (letters). *Congressional Record, 114(No. 46)*: March 20, 1968.

The author defends the formulation of guidelines to cope with the legal, social, and ethical implications involved in research with patients. He says, in part, "the great majority of our medical researchers . . . would still welcome guidelines that would protect us all from the indiscretions of a few."

233. Krause, Axel: Trade secrets: Robert Aries airs his views. *Chemical Engineering, 73*:175-178, April 11, 1966.

The author interviews Robert S. Aries who lists many ways of gathering competitive information and discusses some ethical and legal technicalities involved.

234. Krause, D. C.: Lunar IX pictures: a question of ethics. *Science, 151*: 1477, March 25, 1966.

The author argues that the rush of British and Russian scientists to distribute and comment on the Lunar IX pictures was a violation of scientific ethics: The Russians, who had performed the research, should have had the first opportunity.

235. Labine, R. A.: Engineers ask for action. *Chemical Engineering, 72*: 188+, February 15, 1965.

This article is a report of a survey of professional engineers on professional activism. One part of the survey dealt with an Ethical Review Board and approximately 85% of the respondents favored it.

236. Labine, R. A.: Where is engineering pointing. *Chemical Engineering, 71*:138+, October 26, 1964.

During a discussion of the question of more professionalism in engineering, the author calls attention to the need for "a clearing-house for ethical problems."

237. Lader, L.: Who has the right to live. *Good Housekeeping, 166*:84-85+, June, 1968.

Fewer than 10 percent of the people considered "ideal" for transplants can be saved, due to the lack of availability of equipment and personnel. Lader lists ethical problems of deciding who shall be in the 10 percent, and briefly discusses legal standards of "consent" and "death" and concludes that the great moral issues of transplants can no longer be left to scientists and institutions, but must become a part of public policy.

238. Ladimer, I., and Newman, R. W.: *Clinical Investigation in Medicine: Legal, Ethical, and Moral Aspects.* Law-Medicine Press, 1963.

This anthology covers clinical investigation with regard to the historical and ethical perspective, codifications and principles, legal review and analysis, religious and moral commentary, clinical drug trials, scientific design and technique and research subjects. The volume includes an excellent bibliography on medical ethics.

239. Langer, E.: Court martial of Captain Levy: Medical ethics vs. military law. *Science, 156*:1346-1350, June 9, 1967.

A report not only on the issue of the courtmartial of Captain Howard Levy for refusing to train Special Forces men in dermatology, but also, of the broader issue of the ethical responsibilities and obligations of a physician—and medicine in general—in the armed forces.

240. Langer, Elinor: Human experimentation: New York verdict affirms patients' rights. *Science, 151(No. 3711)*:663, February 11, 1966.

The author states some regulations on medical researchers by Regents of the University of New York. The Regents define "'informed' consent" and how far a physician may exercise his physicians authority when he is acting in the role of experimenter.

241. Lanz, Henry: Letters—'code of ethics.' *Science, 142*:916, November 15, 1963.

The author in reply to L. Cranberg (*Science, 141*:1242, 1963) states that "the game of science is played under certain rules— uncodified, yes, but nevertheless present and adhered to by most scientists," and that "the mere thought of setting up a code of ethics for scientists is insulting."

242. Lauber, Joseph J.: Correspondence—preserving life. *America, 103 (No. 21)*:545; August 20, 1960.

In reply to the editorial, "Patient, Doctor and Human Life," (*America;* July 16, 1960) the author, an M.D., provides further clarification of "ordinary" and "extraordinary" means of preserving life. The issue of transplants is not mentioned.

243. Lear, John: Do we need new rules for experiments on people, *Saturday Review, XLIX(No. 6)*:61-70, February 5, 1966.

The author considers the Southam-Mandel (cancer experimentation) Case in some detail and proposes certain ethical obligations on physicians and legal reforms in civil actions against them.

244. Lear, John: Human guinea pigs and the law. *Saturday Review, 45*: 55-57; October 6, 1962.

The author contends that governmental authority is needed to prevent ethical violations on abuses of personal life and dignity caused by certain drug experiences.

245. Lear, John: Morality in science: report on a crisis. *Saturday Review, 46*:49-54, March 2, 1963.

The author stresses the need for certain ethical guidelines within science.

246. Lear, John: Public policy and the study of man. *Saturday Review,* pp. 59-62, September 7, 1968.

Lear reviews recent developments to create a larger role for behavioral scientists in the federal government.

247. Lear, John: Research in America: experiments on people—the growing debate. *Saturday Review, 49*:41-43, July 2, 1966.

The author outlines the Southam-Mandel case, the new policy of the Public Health Service (that research being funded with public money must have "independent review"), the history and substances of Dr. Henry Beecher's charges of ethical misconduct in experiments, and concludes by noting that the ethical debate involves only experiments which are clearly nontherapeutic. Ethical questions should be dealt with, he argues, with regard to experiments which are, according to the physician doing the experiment, designed to benefit that patient.

248. Lear, John: Struggle for control of drug prescriptions. *Saturday Review, 45*:35-39, March 3, 1962.

The author gives meticulous detail to seven "sets of facts" to show the ethical violations with regard to drugs committed by doctors.

249. Lear, John: Summons to science: apply the human equation. *Saturday Review, 45*:35-39, May 5, 1962.

The author's thesis is that scientists must not forget that science does not operate independently of humans; thus, the human factor is a necessary part of the scientific ethic.

250. Leary, Timothy *et. al.*: The politics of the nervous system. *Bulletin of the Atomic Scientist, XVIII(No. 5)*:26, May, 1962.

In this reply to James Lieberman's "Psychochemicals as Weapons" (*Bulletin of the Atomic Scientist;* January, 1962), Leary (*et. al.*) states that only as a result of ignorance and misinformation, can consciousness-expanding drugs be harmful. He advocates "accurate information, openly shared, and calm, courageous response to the evidence."

251. Lederberg, Joshua: Experimental genetics and human evolution. *Bulletin of the Atomic Scientist, XXII(No. 8)*:4-11, October, 1966.

The author lists reasons why genetics research is now so important. He says the trend is going from eugenics to euphenics, which means treatment of genetic maladjustments earlier in a person's development. He also lists considerations for a genotype and discusses implications of vegetative propagation.

252. Lederberg, Joshua: Experimental genetics and human evolution, *The American Naturalist*, *100(No. 915)*:519-531, September-October, 1966.

This is a discussion of human evolution in a genetic context. The discussion raises ethical and technical problems of eugenics and euphenics as well as present and possible developments.

253. Lee, Louis: The younger viewpoint—social consciousness. *Civil Engineering*, *37*:71, February, 1967.

The author feels that professional engineers are not fulfilling their responsibilities to the public. He recommends a more rounded education and a professional program of community involvement for engineers.

254. Leitenberg, Milton: Science bookshelf—science and man's fate on his planet. *Scientific Research*, pp. 89-92, October, 1967.

The author, scientific director of the Committee for Environmental Information, reviews Barry Commoner's *Science and Survival*.

255. Lessels, G. A.: Stepping-stones to professionalism. *Chemical Engineering*, *71*:86-90, August 31, 1964.

The author argues that the basic "stepping-stone" of the professionalization of engineering is "a highly developed sense of ethics."

256. Levi, L.: Science: a force for unity. *Science News Letter*, *81*:346-347, June 2, 1962.

This article discusses international cooperation among scientists.

257. Levy, L.: Scientists enter politics. *Science News Letter*, *78*:106-107, August 13, 1960.

This article discusses political activities and endorsements of scientists.

258. Lewis, Robert: Letters to the Science Editor—Medical Ethics. *Saturday Review*, *49*:52, August 6, 1966.

Dr. Lewis contends that whenever a university undertakes an experiment and asks subjects to participate, it "guarantees" "modern scientific treatment of professed human information" which constitutes an element of medical ethic. Such a guarantee cannot come from one man or department, but "only from an institution."

259. Libby, W. F.: Mankind's adjustment to scientific advances. *Science Digest*, *50*:84+, October, 1961.

The author traces the relationships between science and society.

260. Lieberman, E. James: Psychochemicals as weapons. *Bulletin of the Atomic Scientist*, *XVIII(No. 1)*:11-14, January, 1962.

Dr. Lieberman lists certain psychochemicals under consideration about 1960 for war purposes and considers some of the physiological and ethical dangers of such use.

261. Lieberman, E. James: The ethical neutrality of LSD. *Bulletin of the Atomic Scientist, XVIII(No. 6)*:41; June, 1962.

In this reply to Leary's (*et. al.*) reply (May, 1962) to Lieberman's "Psychochemicals as Weapons" (January, 1962), Lieberman attacks Leary's defense of LSD and reiterates the danger from LSD and other much more toxic psycho-chemicals.

262. Logan, Donna: Liver advances dramatic. *The Denver Post-Bonus*, p. 3, September, 1968.

The author documents advances in liver transplants and then considers transplantation of other organs.

263. Logan, Donna: New medicine sires agonizing queries. *The Denver Post-Bonus*, pp. 5-6, September, 1968.

The author interviews religious authorities concerning the ethical questions raised by human organ transplants.

264. Logan, Donna: Transplants moral issue for doctors. *The Denver Post-Bonus*, p. 7, September, 1968.

The author interviews surgeons concerning ethical questions in organ transplants. The major part of the article deals with the criteria of death.

265. Logan, Donna: Transplants: right or wrong. *The Denver Post-Bonus*, pp. 1-2, September, 1968.

The author interviews Colorado surgeons, lawyers, and governmental officials concerning the following questions. Donor organ consent; who gives it? Priority; who will receive it? Death; what is the legal definition? And doctors; which ones are qualified?

266. Lonsdale, Dame Kathleen: Science and ethics. *Nature, 193(No. 4812)*: 209-214, January 20, 1962.

The author discusses the nature of Science and the scientist, the nature of ethics as a "system or doctrine of morality," the question of the "responsibility" of the scientist, and, finally, specific moral considerations and decisions that must be made by the individual scientist as well as by the policy-planning public.

267. Loomis, F.: Who shall be the judge. Condensed from Consultation Room. *Reader's Digest, 86*:91-94, April, 1965.

The author, in this emotional drama, argues that a doctor must continue to fight for a patient's life until there is no hope left.

268. Lovell, Bernard: Lunar IX pictures: a question of ethics. *Science, 151*:1477, March 25, 1966.

In reply to Dale Krause's charge that the Jodrell Bank Observatory "rushed" to distribute the Lunar IX pictures, Lovell states that the British and American scientists did not receive the photograph until after the Russians had held a press conference.

269. Lowe, George E.: The Camelot affair. *Bulletin of the Atomic Scientist, XXII(No. 5)*:44-48, May, 1966.

This gave the history of Project Camelot and the involvement of various governmental departments. Lowe concludes that even if the government agencies (State, DOD, CIA) were kept out, all American sponsored social science research will be suspect of frustrating social change.

270. Maccoby, Michael: Social psychology of deterrence. *Bulletin of the Atomic Scientist, XVII(No. 7)*:278-281, September, 1961.

Maccoby, in attacking the "game theory" of deterrence uses what are, at least by implication, ethical arguments. In doing so, he puts certain ethical limits on science and technology.

271. Macleish, Archibald: To face the real crisis: man himself. *New York Times Magazine*:5+, December 25, 1960.

The author suggests that the real crisis man must face is not some external threat but, actually, man himself.

272. Macleod, Kenneth I. E.: Correspondence—attitude toward human experimentation. *New England Journal of Medicine, 274(No. 14)*: 791; October 6, 1966.

The author, Commissioner of Health, in reply to William Curran's The law and human experimentation, (*New England Journal of Medicine*, August 11, 1966), states that, "we had better take a hard look at some of the things we do in the name of medical science," and that the informed consent of a patient should always be obtained.

273. Mandell, Loring: Letters. *Bulletin of the Atomic Scientist, XXII(No. 4)*: 31, April, 1966.

In part, this is a reply to Dash's January, 1966 article. Mandell claims he misused a quote and oversimplified.

274. Mann, Kenneth W., Rev.: Hearings to resume on health science commission—exhibit 1 (letter). *Congressional Record, 114(No. 46)*, March 20, 1968.

The author, Executive Secretary, Division of Pastoral Services of the Episcopal Church, supports the proposal for a Commission on Health, Science and Society.

275. Margenau, H.: *Open Vistas*. New Haven, Yale University Press, 1961.

The author divides this book into the following areas: science and human affairs; the inner light of reason; esthetics and relativity; and the decay of materialism in our time; the fade-out of concrete models, and reality, determinism, and human freedom.

276. Margolis, H.: Consultants and conflicts. *Science, 135*:88-89, January 12, 1962.

The author discusses some attention-getting problems of the scientist as consultant.

277. Margolis, H.: Science advisory committee and national goals reports emphasize growing roles of government. *Science, 132*:1648-1649, December 2, 1960.

The author considers the increasing interaction between science and government.

278. Margolis, H.: Scientific advisers. *Science, 134*:1739, December 1, 1961.

The author discusses certain problems in the present (1961) system of "scientific advisers" but claims no new system can be offered.

279. Marley, Faye: Are human tests ethical? *Science News, 90*:115, August 20, 1966.

This article reports a grant by the U. S. Public Health Association to study ethics of human experimentation, and reports on the charges of Dr. Henry Beecher, the Declaration of Helsinki, and the Nuremburg Code.

280. Marlowe, D. E.: Legacy of Merlin. *Mechanical Engineering, 86*:26-29, February, 1964.

Merlin was the magician of King Arthur's domain. The king regularly sought his advice. Today, the author metaphorizes, the magician is the engineer but they are not regularly sought. "Engineers . . . must find the way to make their knowledge available to government bodies.

281. Meade, J. E., and Parkes, A. S. (Editors): *Biological Aspects of Social Problems.* Plenum Press, New York, 1965.

This book is a report of a symposium on biological aspects of social problems, organized by the Eugenics Society and held in October, 1964 at University College in London. Areas covered are: population trends; social mobility and education; genetic aspects of medicine, and aspects of fertility control.

282. Measuring engineering efficiency. *Chemical Engineering, 69*:91-92, December 24, 1962.

The author discusses a system of Performance and Cost Evaluation (PACE) which can measure engineering efficiency but which has been termed unethical by some because, say the critics, it is "nothing more than old-fashioned spying."

283. Medical Ethics Debate Boils. *Science News, 93*:282-283, March 23, 1968.

This news article considers testimony before Senator Mondale's Health Science proposed commission and notes public and individual professional's attitudes towards medical ethics in general, and heart transplants in specific.

284. Meyer, Herbert M.: The beginning of the common-sense. *Bulletin of the Atomic Scientist, XXII(No. 2)*:23-25, February, 1966.

Scientists have, for the most part, ignored ethical questions and such attempts to set up ethical codes as there have been have failed because of an inability to communicate "common denominators" of ethical values.

Because of scientific advances in which "we are not even aware that we destroy social orders all over the globe without providing alternatives," scientists must now grapple with basic ethical issues. Meyer offers a "new three-dimensional set of co-ordinates:" personal life; scientific working life, and life within our society. Meyer calls for reaching the young (high school, college, and graduate students) to become aware of the need for ethics.

285. Miller, Arthur Selwyn: Experiments on humans—where are the lawyers? *Saturday Review, 49*:48-50, July 2, 1966.

The author argues that science and technology have risen to the point where they are so powerful that external controls ought to be instituted. Logically, lawyers should play a major role in establishing society-wide controls, but the lawyers have not seen this problem and have defaulted in their responsibility.

286. Miller, Cecil: Human living—codes and problems. *Consulting Engineer,* pp. 110-114, February, 1967.

The author states that, "if anywhere," the keys to the problems of human living may be found in "the funded experience of professional organizations, especially as it finds expression in their various codes of ethics."

At the conclusion of his article, he states that answers to questions of morality and, by implication, to particular moral questions (e.g. ethics of the scientist or researcher) "are to be found in rededication to the homely precepts of Hippocrates."

287. M. I. T. and the selective service. *Bulletin of the Atomic Scientist, XXIV(No. 3)*:35, March, 1968.

A statement issued from the public relations office of M. I. T. on January 18, said, in part, that science students should not be given preference, in the granting of graduate deferments, over non-science students, "in the absence of a broad national emergency."

288. Mitchell, George: Heart association president probes transplant ethics. *Minnesota Daily,* reprinted in *Congressional Record, 114(No. 139),* March 11, 1968.

The author interviews Dr. Jesse E. Edwards, President of the American Heart Association, concerning the ethical problems involved in heart transplants.

289. Modell, W.: Hazards of new drugs. *Science, 139*:1180-1185, March 22, 1963.

The author holds that there are certainly ethical obligations to the subject of drug experiments and he adds that there are also ethical obligations to the nonsubject—"to the patient who will receive the new drugs in clinical practice."

Thus, he argues, many dilemmas relative to the merits of drugs, can be avoided if the ethical rules are observed, "If the experiments which lead to clinical use are ironbound, if the publication of results is withheld until the proof is in, and if general use is not initiated and pressed until the critical questions have been decisively answered through extensive trial in clinical practice."

290. Mondale, Walter: Hearings to resume on health science commission. *Congressional Record, 114(No. 46),* March 20, 1968.

The author discusses the importance of the proposed commission and discusses the testimony of Drs. Najarian, Kantrowitz, Beecher, Lederberg, Barnard and others.

291. Mondale, Walter: Introduction of joint resolution to establish a commission on health science and society. *Congressional Record, 114 (No. 19),* February 8, 1968.

The author and fifteen co-sponsors wish to establish a commission which would "undertake a comprehensive investigation and study of the legal, social, and ethical implications of health science research and development . . ." Evidence citing the need for analyses of these areas is given.

292. Moore, Francis D., M.D.: Ethics in new medicine: tissue transplants, *Nation, 200:*358-362, April 5, 1965.

Dr. Moore lists four guidelines for experimentation: (that the patient and family understand alternatives available and that the patient enter into the procedure of his own free will; that each patient must receive the best and most experienced medical care available; that preliminary laboratory study justifies the attempt, and that each patient's case be studied and documented as carefully as possible and be made available to the general view). He also states three specific principles for liver transplants: (every effort is made to assure maximum donor-recipient tissue compatability; both kidneys of the donor are normal; and the donor understands the alternatives, risks, and uncertainties).

293. Morrison, P.: Where is science taking us. *Saturday Review, 45:*46, July 7, 1962.

The author lists implications of certain recent developments in science.

294. Moss, John E.: The crisis of secrecy. *Bulletin of the Atomic Scientist, XVII(No. 1):*8-11+, January, 1961.

Congressman Moss claims that the vast and improper use of classified labels to withhold documents and studies from public examination is ethically improper.

295. Muller, H. J.: Science for humanity. *Bulletin of the Atomic Scientist,* 15(No. 4):146-150, April, 1959.

The author discusses the role of science in relation to society.

296. Muller, H. J.: The meaning of freedom. *Bulletin of the Atomic Scientist,* XVI(No. 8):311-316, October, 1960.

The ethic of modern science "demands the utmost freedom and frankness of communication, criticism, and countercriticism by all engaged in it." The issues involved in scientific progress cannot be passed upon by, "politicians, lawyers, leaders of governments, religions, or other ideologies."

297. McDonald, Donald: Scientist as citizen. *Bulletin of the Atomic Scientist,* XVIII(No. 6):25-28, June, 1962.

During the course of this interview by McDonald with Hans Bethe, Dr. Bethe discusses some scientists' response to ethical problems.

298. McGee, Daniel B.: Hearings to resume on health science commission—exhibit 1 (letter). *Congressional Record,* 114(No. 46), March 20, 1968.

The author, Associate Professor of Christian Ethics at Baylor University, defends the purpose and practical need for the commission on science and health research called for by Senator Walter F. Mondale.

299. McNamara, Raymond W.: The younger viewpoint—code of ethics. *Civil Engineering,* 37:71, February, 1967.

The author reports on the role of the American Society of Civil Engineer's enforcement and maintenance of its Code of Ethics.

300. Nader, Claire: The technical expert in a democracy. *Bulletin of the Atomic Scientist,* XXII(No. 5):28-30, May, 1966.

At present, as exemplified in the issue of water fluoridation, the involvement of scientists in public policy "indicates their lack of understanding of a scientific issue in its socio-political context."

Nader advocates an "ethic of responsibility" to devise a mechanism to achieve consensus of common interests and standards of implementation and to force conflicting interests to a public hearing.

301. Nelson, B.: Anthropologists overwhelmingly approve research ethics statement. *Science,* 156:365, April 21, 1967.

The "Statement on Problems of Anthropological Research and Ethics" adopted by the Fellows of the American Anthropological Association in April, 1967, in part, deplores anthropological research as a cover for foreign intelligence activities; holds that universities, except in wartime, should not take contracts in anthropology "not related to their normal functions!" and says that anthropologists have an ethical duty to "decline to participate in or accept support from organizations that permit misinterpretation of technical competence, excessive costs, or concealed sponsorship of activities."

302. Nelson, James B.: Hearings to resume on health science commission—exhibit 1 (letter). *Congressional Record, 114(No. 46),* March 20, 1968.

The author, of the United Theological Seminary of the Twin Cities, has become increasingly convinced of the need for careful, cross-disciplinary, social reflection "upon issues of technological advance, so that guide lines may be evolved that benefit the whole of society."

303. New administration: it faces a number of questions of scientific policy: no easy solutions in sight. *Science, 132:*1382-1383, November 11, 1960.

This article tests many of the problems of the science policy to be faced by the incoming administration.

304. Nicholson, E. K., and Gammell, John: Ethics and the technical societies. *Electrical Engineering, 81:*260-261, April, 1962.

The authors list procedures for the reviewing and updating of codes of ethics by the professional engineering societies.

305. Oppenheimer, J. Robert: In the keeping of unreason. *Bulletin of the Atomic Scientist, XVI(No. 1):*18-22, January, 1960.

Oppenheimer's point is that, in the present civilization, the society has abdicated its responsibility for philosophical and ethical discourse. He seems to imply that the uses of science and technology should be regulated somehow by this ethical discourse.

306. Osborne, Burl: I didn't think about dying. *The Denver Post-Bonus,* p. 8, September, 1968.

The author, the recipient of a transplanted liver, discusses the question of transplants from a personal point of view.

307. Page, I. H.: Medical ethics. *Science, 153:*371, July 22, 1966.

The author holds that, in the absence of "an expert in medical ethics," the spirit of the words of the Hippocratic Oath should "provide a beacon for both today and tomorrow" for all concerned with medical ethics. By implication, the question is raised whether individual practitioners or the medical profession in general can face ethical problems in more than an individual and *ad hoc* manner.

308. Pandullo, Francis: Ethics and municipal engineers in private practice. *American Society of Civil Engineers Journal of Professional Practice, 91(No. 2-4455):*1-6, September, 1965.

The author discusses four types of relationships of the engineer in private practice that may present ethical problems.

309. Panel on Privacy and Behavioral Research: Privacy and behavioral research. *Science, 155(No. 3762):*535-538, 1967.

In this preliminary summary, the panel (appointed by the Presidest's Office of Science and Technology, in January, 1966) concludes in part, that participation by subjects in experiments must be voluntary and "based on informed consent to the extent that this is consistent with the objectives of the research;" "the scientist has an obligation to insure no permanent physical or psychological harm to the subject; the scientist must protect the privacy of the subject both in the research and in the published reports; and that legislation to insure recognition of human rights of subjects is "neither necessary nor desirable."

310. Park, I. R.: Are engineers too good for politics? *Product Engineering, 31*:24-25, November 7, 1960.

This article presents points of view concerning political activity by professional engineers.

311. Pathologists—antitrust and ethics. *Time*, p. 52, July 15, 1966.

This article reports a suit filed on the College of American Pathologists by the Justice Department charging "price-fixing" and other illegal actions. The suit asks the Court to prescribe a "new set of business ethics" for the pathologists.

312. Patient, doctor, human life. *America, 103*:451, July 16, 1960.

This editorial states as the position of the Catholic Church that, briefly, persons are required to take ordinary, but not extraordinary means to preserve life. Some examples of each are given. A doctor, however, has stricter duties in that "he must not only do the minimum to which the patient is bound, but also do whatever the patient reasonably requests as well as what professional standards require."

313. Pauling, Linus: Peace on earth: the position of the scientists. *Bulletin of the Atomic Scientist, XXIII(No. 8)*:46-48, October, 1967.

The scientists must educate the public about the threat from nuclear and other weapons. "It is imperative that, without delay, an international agreement be made to stop further R & D of these frightening and immoral methods of mass murder."

Science does not reject morality: it is forced to accept the Golden Rule.

The United States must end the war in Vietnam.

"We must not destroy this world. We must not destroy the human race."

314. Pesin, Edward, and Winter, Ruth: Organ transplants: a legal and moral dilemma. *Science Digest, 63*:68-72, April, 1968.

The authors raise legal and moral perplexities involved in heart and other organ transplants.

315. Piel, Gerald: Federal Funds and science education. *Bulletin of the Atomic Scientist, XXII(No. 5)*:10-15, May, 1966.

Piel's thesis is that the university must be considered as "vessels that cherish and enlarge" liberties of citizens and not as "instruments of national purpose." Thus, federal funding must be concerned first, with making the universities "autonomous centers of creative initiative" in American life.

316. Pierce, J. R.: Freedom in research. *Science, 130*:540-542, September 4, 1959.

The author argues for the right of the scientist to engage in research without undue interference.

317. Pierce, J. R.: The paper dragon . . . a tale of the times. *Physics Today, 16(No. 8)*:45-50, August, 1963.

The author uses satire and irony to depict the dishonesty of what might be termed the scientific community; academic, governmental, and private.

318. Plant, M. L.: Informed consent—new area of malpractice liability? In: *Medical Malpractice.* Ann Arbor, Michigan, Institute of Continued Legal Education, 21-43.

This book is comprised of lectures dealing with ethical and legal issues relating to medical malpractice. One section deals especially with the question of "informed consent."

319. Post, R. H.: Eugenics and the I.U.C.D.'s. *Eugenics Quarterly, 12 (No. 2)*:112-113, 1965.

It is suggested that a private agency research all cases of childbirth resulting from malfunction of I.U.C.D.'s as a prelude to the study of the role of genetics, natural selection, and eventually eugenics in the production of NUCD (Nitra-uterine) gestated individuals.

320. Powledge, Fred: What will the doctors do for Jean Paul Getty that they won't do for you? *Esquire,* pp. 23-27+, October, 1968.

The author details the inequality of medical care for the rich and for the poor and raises ethical questions about same.

321. Protecting human guinea pigs. *Business Week,* p. 71, July 23, 1966.

A news report on the Public Health Service's rules for research grants.

322. Pryor, W. J.: Are medical ethics an anachronism. *New Zealand Medical Journal, 62(No. 369)*:203-206, 1964.

The author presents a history of medical ethics, discusses the relationship between ethics and the layman, and, finally, suggests that the medical profession review its ethics for applicability from time to time—to keep medical ethics from becoming an anachronism.

323. Rabinowitch, Eugene: Responsibilities of scientists in the atomic age. *Bulletin of the Atomic Scientist, XV(No. 1)*:2-7, January, 1959.

The primary ethical responsibility of the scientist is to educate the public and its leaders of the uses of science and technology. This implies scientific investigation of any given area. Rabinowitch takes a peculiar position on the action of the scientist (i.e. refusal to work for military purposes; collective action to stop arms race, etc.). Apparently *individual* decisions aren't necessary because they are ineffective; although if, somehow, some collective action (that could be effective) began to take place, it seems to be ethically proper.

324. Randal, Judith: Hearings to resume on health science commission—exhibit 1: "Naive howls on medical research." *Congressional Record, 114(No. 46)*, March 20, 1968.

The author claims that "who should get what medical treatment" ought to be examined by people other than doctors as well as the medical profession itself.

325. Rapport, S. (Editor): *Science: Method and Meaning.* New York, New York University Press, 1963.

This book of readings is divided into two sections; first, science and the scientist (the nature of scientific activity, observation, hypothesis, experiment, and discovery), and second, science and "the world around us."

326. Research and responsibility, (Editorial). *Nation, 202*:284-285, March 14, 1966.

This editorial apparently supports the decision of the New York Regents in the Southam-Mandel Case and claims that "no fact may be concealed" from a subject in an experiment. The subject is the only person who will decide what information is relevant and what isn't.

327. Rosenfeld, A.: Search for an ethic. *Life, 64*:75-76+, April 5, 1968.

The author considers when a doctor may experiment and raises the question of whether, in certain circumstances, doctors have a moral right not to transplant. Also discussed are definitions of death and legal aspects of acquiring cadaver organs.

328. Rowe, R. R.: Ethics in public practice. *Civil Engineering, 29*:1-2, January, 1959.

The author discusses the applicability of the Code of Ethics of the American Society of Civil Engineers to the engineer in private practice article by article.

329. Rowley, Louis N.: In the public interest. *Mechanical Engineering, 90(No. 8)*:14-16, August, 1968.

The author, president of the American Society of Mechanical Engineers, argues that it is the professional duty of the engineer to assume a role in national planning and to mediate between and bring together public and private interests.

330. Royal Medico—Psychological Association: The Royal Medico—Psycho-

logical Association's Memorandum on Therapeutic Abortion. *British Journal of Psychiatry, 112(No. 491)*:1071-1073, 1966.

This memorandum opposes making abortion lawful on the grounds of inconvenience to the parents and suggests that abortion should follow the rules of other standard medical procedures when two doctors agree that it is medically proper.

331. Ruebhausen, O. M., and Brim, O. G., Jr.: Privacy and behavioral research. *American Psychologist, 21*:423, 1966.

The authors discuss the moral claim to private personality, the nature of privacy, the scientific challenge, the need for equilibrium, behavioral research and individual privacy, the concept of consent, the concept of confidentiality, and, finally, argue for the adoption of a code of ethics for behavioral research.

332. Russell, Bertrand: The social responsibilities of the scientist. *Science,* February 12, 1960.

The scientist must be responsible for the uses that are made of his knowledge.

333. Russell, J. E.: Proper use of common sense and engineering in secondary recovery. *Journal of Petroleum Technology, XVI*:1003-1005, September, 1964.

The author stresses the need for a formal ethic in the engineering profession as a basis for uplifting the image of the engineer.

334. Rzasa, M. J.: Ethics is a personal thing. *Chemical Engineering Progress, 61*:35-37, April, 1965.

The author holds that a code of ethics is necessary for engineering to be considered a profession. In addition, he recommends, for engineering students: a course in engineering professionalism, to include ethics; reference, within the technical curriculum, to ethical practices; and professional and student seminars on engineering ethics.

335. St. John-Stevas, Norman: *Life, Death, and the Law.* Meridian Books, 1964.

The author discusses legal and ethical problems of certain scientific and medical developments.

336. Schwab, R. S.; Potts, F., and Bonazzi: EEG as an aid in determining death in the presence of cardiac activity. *Electroencephalography and Clinical Neurophysiology, XV(No. 1)*:147-148, 1963.

In this abstract, the authors offer the following seven criteria for establishing death: absence of spontaneous respiration for thirty minutes; no tendon reflexes; dilated pupils and no pupillary reflexes; no change in heart rate with eyeball pressure; no EEG activity for thirty minutes; no EEG discharge when a loud noise is made; and last, an inter-electrode resistance of usually over 50,000 Ω. "With this information before him, the physician can then establish the time of death and order the cessation of respirators and other automatic equipment."

337. Scientists urged to assume greater social responsibility. *Science Digest,* *48,* inside back cover, November, 1960.

This article presents certain social responsibilities and ethical considerations for scientists.

338. Scott, James L., M.D.; Belkin, Gerald A., M.D.; Finegold, Sydney M., M.D., and Lawrence, John S., M.D.: Correspondence—human experimentation. *New England Journal of Medicine, 275(No. 14):* 790-791, October 6, 1966.

The authors, in reply to Henry Beecher's "Ethics and Clinical Research," (*New England Journal of Medicine;* June 16, 1966) charge that "Dr. Beecher quotes out of context, oversimplifies and otherwise distorts the purpose and findings of our investigation of the hematologic toxicity of chloramphenicol, reported in the *New England Journal of Medicine, 272*:1137, 1965."

339. Sears, Paul B.: Man and his habitat: the perspective of time. *Bulletin of the Atomic Scientist, XVII(No. 8):*322-326, October, 1961.

A blind increase in population and industrial civilization threatens the ecological balance, and human dignity. In addition, our society's value of science and technology is so high and so "verging on fear" that we forget that science ought to be shaped and based upon our values.

340. Shapiro, Ivan: Medical experiments. *New Republic, 156(No. 4):*37-38, January 28, 1967.

In this reply to M. Alderman's article in the December 3, 1966 issue, Shapiro argues that all scientific experimentation performed by doctors upon persons which is done without the "full knowing consent" of the patient, and for the benefit of the patient, probably constitutes malpractice.

341. Should doctors ever let patients die. *Christian Century, 79*:857, July 11, 1962.

This editorial states that mercy killing is still killing, but certain measures which could prolong life need not be taken, and, in fact, "to prevent the coming of that death is to commit an unmerciful sin."

342. Sidel, V. W.: Medical ethics and the cold war. *Nation, 191*:325-327, October 29, 1960.

The author deals with the question of the confidentiality of medical records of persons whose actions may be of consequence to national security. In particular, he considers the case of William H. Martin, who defected from the National Security Agency in 1960.

343. Silverman, William A.: Correspondence—human experimentation.

New England Journal of Medicine, 275(No. 14):790, October 6, 1966.

The author, in reply to Henry Beecher's "Ethics and Clinical Research" (*New England Journal of Medicine;* June 16, 1966), suggests that Beecher concentrated on the "sins of Commission," of human experimentation; but that they do not compare with "evils of failure to conduct human investigation," which predominate.

344. Sinsheimer, Robert: The end of the beginning. *Bulletin of the Atomic Scientist, XXIII(No. 2)*:8-12, February, 1967.

The consequences of molecular biology—changes in life span, size, sexuality, diseases, hunger, intellectual range and capacity, density of man on this planet—will, because the above form and underlie our society, all but wreck our society. Are ethics responsible for these changes? "We must seek to plan a balance between change and order;" "changes must be orderly and with humanity aforethought."

345. Slater, Carl H.: Letters to the science editor—medical ethics. *Saturday Review, 49*:51, September 3, 1966.

The author, a medical student, holds that an open discussion of medical ethics is beneficial because, "not a month goes by that I do not find something in the medical journals to make me question the ethical wisdom of my superiors."

346. Snow, C. P.: Moral un-neutrality of science. *Science Digest, 49*:19-24, March, 1961.

Snow says that he rejects the argument that science is morally neutral but nowhere in the rest of the article does he defend an alternative position. The author finally states that it is the moral duty of the scientist to explain the "either-or" of any contemplated course of action.

347. Sonneborn, T. M. (Editor): *Control of Human Heredity and Evolution.* Macmillan, New York, 1965.

This book deals with many ethical questions concerning recent biological advances. In the symposium are: S. E. Luria, Edward L. Tatum, Robert De Mars, G. Pontecorvo, and Herman J. Muller.

348. Spitzer, W. O. Are heart transplant moral. *Christianity Today, 12*: 24-26, February 16, 1968.

The author raises and discusses the following ethical concerns: Are criteria of death acceptable to society as well as doctors? Is cardiac transplantation experimental or truly therapeutic? Are the recipients really being helped? Who decides who shall live and who shall die?

349. Steinbach, H. Burr: Scientists and public policy. *Bulletin of the Atomic Scientist, XVIII(No. 3)*:10-13, March, 1962.

The author discusses scientists role in the formation of public policy at the present (1959); what that role should be, and what steps would be necessary to bring that role about.

350. Stewart, Bruce: Science and social change. *Bulletin of the Atomic Scientist, XVII(No. 7)*:267-270+, September, 1961.

Scientists have an ethical responsibility to bring to society "those mental qualities and attitudes which have made him successful in science: critical analysis" leading to constant revision of old theories and search for new ones, and demand for evidence at the expense of tradition.

351. Stewart, William H., M.D.: An invitation to open dialogue. *Saturday Review, 49*:43-44, July 2, 1966.

Dr. Stewart, the Surgeon-General of the Public Health Service, claims that existing law is firm and clear on the principle that the decision to become a subject for research must be made by that subject. He then states that, because of this, the PHS is "asking that the institution assure us that research proposals related to the use of human subjects are being systematically subjected to independent review, and we are urging that qualified individuals from outside the scientific area be involved in this review.

Dr. Stewart concludes by stating his awareness that "some important research will be delayed or perhaps lost" because of the new regulations and invites a "continuing dialogue" to evolve better solutions. This article is an excerpt of his talk (April 30, 1966) to the American Federation for Clinical Research at Atlantic City.

352. Strauss, Anselm L.: Medical ghettos. *Trans-Action, IV(No. 6)*:7-15+, May, 1967.

The author argues that radical transformation of medical service as it presently stands will be needed to provide equal care to the poor. He details the "second-rate" medical care the poor now receive and lists specific recommendations.

353. Stucki, Jacob C.: Letters—experimentation: rights and risks. *Science, 155(No. 3770)*:1617, March 31, 1967.

In this reply to Wolf Wolfsenberger's "Ethical Issues in Research with Human Subjects" (*Science*, January 6, 1967), Stucki claims that he did not raise enough ethical questions and arguments relative to "experimental activities and procedures employed but not consciously recognized or formally labeled as research."

354. Szent-Gyorgyi, A.: Brain, morals, and politics. *Bulletin of the Atomic Scientist, 20*:2-3, May, 1964.

The author argues that the world has changed in such a way as to change the nature of our moral code. Thus, scientists must look at all the people in the world as one (not as differing groups) with himself and does not have the right to manufacture atom bombs to kill them.

355. Tangerman, E. J.: How secure can you keep your design secrets. *Product Engineering*, 37:87-95, July 4, 1966.

The author is concerned with design secrets of industry. His distinction between "ethical" and "unethical" spying is humorous and, presumably, tongue-in-cheek.

356. Taylor, Carl E.: Ethics for an international health profession, *Science*, 153(No. 3737):716-720, August 12, 1966.

In public health, the "patient" is not one person but a whole population unit. Thus, it is necessary that the PH doctor be responsible for many decisions balancing costs and benefits in economic terms.

In international public health, Taylor calls for, first, a sharing of information among colleageus and, second, a constant awareness of social conditions in other countries.

357. Taylor, G. R.: *The Biological Time Bomb*. London, Thames and Hudson, 1968.

This book makes the bomb look like a toy compared to what is going on in the biosciences. He also suggests and implies where ethical questions have arisen and will soon arise.

358. Teilhard De Chardin, Pierre: *The Future of Man*. New York, Harper & Row, 1964.

The author deals with many ethical problems of science in this book.

359. Tenery, Robert M.: Medical ethics—medical etiquette. *American Medical Association Journal*, 195(No. 13):1137-1138, March 28, 1966.

The author discusses the use of a formula consisting of "three measuring codes" (measure of intent; measure of local custom and laws; measure of the golden rule) to apply to ethical problems, and then considers, in some detail, certain ethical aspects of "professional courtesy."

360. Vallance, Theodore: Project Camelot: an interim postlude. *American Psychologist*, 21(No. 5):441-444, May, 1966.

The author gives an account of the goals and expectations initially hoped for from Project Camelot.

361. Vaux, K.: Heart transplant: ethical dimensions. *Christian Century*, 85:353-356, March 20, 1968.

The author discusses the following three ethical problems: time and meaning of death; the question of donor and recipient, and the rejection phenomena.

362. Viorst, Milton, and Reistrup, J. V.: Radon daughters and the federal

government. *Bulletin of the Atomic Scientist, XXIII(No. 8)*:25-29, October, 1967.

The federal government has not done much to protect uranium miners. When the Department of Labor and the Department of Welfare finally took action to demand that mines have a 0.3WL (Working Level) of radiation, they quickly "backed down" under pressure from mine owners and certain congressmen.

363. Visscher, Maurice B.: Medical research and ethics. *American Medical Association Journal, 199(No. 9)*:631-636, 1967.

The author analyzes different moral positions with regard to vivisection and animal experimentation.

364. Walsh, John: Foreign affairs research: review process rises on ruins of Camelot. *Science, 150*:1429-1431, December 10, 1965.

The author documents his thesis that the after-effect of Project Camelot was to be a better system of review of projects in the behavioral sciences undertaken by universities and professional societies for certain governmental agencies.

365. Ward, Leo R. (Editor): *Ethics and the Social Sciences*. South Bend, Indiana. University of Notre Dame Press, 1959.

The following contributors present articles dealing with ethics and the social sciences: Francis G. Wilson, Kenneth Johnston, and James R. Brown.

366. Warwick, Warren J.: Organ transplants: a modest proposal. *The Wall Street Journal*, June 24, 1968.

The author emphasizes many of the ethical problems involved in organ transplants in satirizing them.

367. Washington News—hospital integration guidelines. *American Medical Association Journal, 195(No. 13)*:22-23, March 28, 1966.

This is a report of guidelines for hospitals issued by the federal government to help them comply with the Civil Rights Act. Some of the guidelines are, briefly: that all patients are assigned to all rooms, wards, floors, etc. without regard to race, color, or national origin; that granting of staff privileges is carried out in a non-discriminatory manner; that nondiscriminatory practices include all aspects of training programs and that recruiting be held at both predominantly white and predominantly Negro schools; and that hospitals recently changed from discriminatory practices take steps to notify those who had previously been excluded.

368. Watson-Watt, Robert: Physicist and politician. *Bulletin of the Atomic Scientist, XV(No. 7)*:298-301, September, 1959.

A scientist has, in addition to his general duty as a citizen, a special duty to citizenship. That duty is to clearly express his view of the effects, or possible effects, of the applied use of his scientific research.

369. Weaver, Warren: The moral un-neutrality of science. *Science, 133 (No. 3448)*:255-256, January 27, 1961.

In this introduction of C. P. Snow, the author claims that the ethical decisions made by the scientist affect all other aspects of human life.

370. Webb, Paul: Letters—"experimentation: rights and risks." *Science, 155(No. 3770)*:1617, March 31, 1967.

In this reply to Wolf Wolfensberger's "Ethical Issues in Research with Human Subjects" (*Science,* January 6, 1967), Webb states that it is the duty of the experimenter to expose himself to all or more of the risks to any individual subject.

371. Weidenbaum, Murray L.: A matter for the public to decide? *Bulletin of the Atomic Scientist, XXIV(No. 6)*:7, June, 1968.

In a short editorial, Weidenbaum offers a third position to those of "scientific theologians or wistful yearners for a simpler society": We will not try to stifle scientific inquiry nor inhibit technological innovation. Also . . . the determination of the uses to which public resources . . . are put is a matter for the public to decide."

372. Weigent, C. E., M.D.: Hearings to resume on health science commission—exhibit 1 (letter). *Congressional Record, 114(No. 46)*, March 20, 1968.

The author, pathologist at VA Hospital, supports a commission to study the "legal, social, and ethical issues" of medical research.

373. Weinberg, Alvin M.: Science, choice, and human values. *Bulletin of the Atomic Scientist, XXII(No. 4)*:8-13, April, 1966.

Beginning with advocating a "mission-oriented" system of funding scientific projects, Weinberg then goes on to a "new ethical principle for science: not only must science seek truth, it must seek relatedness."

After reviewing the views of a) Bronowski, Rapoport, Polyani, and b) Bruce Lindsay, he bases his scientific ethic on a combination of them (truth; entropy)—"on the notion that truth is whole, that the purpose of science is not merely to unearth the facts but also to show the relatedness of facts."

Thus, "we decide on the good (of a sphere or universe of activity) from the standpoint of the neighboring universes; in making the judgment, we ask if the activity or attitude we are judging helps create a unity, a harmony in the universes doing the judging. What we are judging is good to the extent to which the answer is yes."

374. Wessel, Morris A.: Correspondence—human experimentation. *New England Journal of Medicine, 275(No. 14)*:790, October 6, 1966.

In reply to Henry Beecher's, "Ethics and Clinical Research," the author supports Beecher's position and argues that medical physicians and experimenters have a primary moral duty to "make use of all available scientific knowledge and skill to diagnose and cure, if possible," and "above all" offer comfort.

375. What the code of ethics says. *Chemical Engineering Progress, 61*:39, February, 1965.

The official code of ethics of chemical engineers is reprinted here.

376. Wheeler, K., and Lambert, W.: Uneasy balance, ethics vs. profits; physicians who profit from prescribed medications. *Life, 60*:86-88+, June 24, 1966.

Because the patient accepts a doctor on faith, he is unusually vulnerable to unethical and improper treatment and prescription of unneeded or possibly harmful drugs.

The authors apparently support drug control legislation (which has since been passed).

377. Whittet, T. D.: Professional ethics, a British view. *West African Pharmacist, 9(No. 3)*:46-51, 1967.

Codes of ethics in pharmacy and hospital practice since that of the Guild of Pepperers are reviewed.

378. Wisely, W. H.: Administration of Ethical Standards. *Civil Engineering, 37*:37, August, 1967.

He charges that the American Society of Civil Engineers is "lackadaisical or ineffective or too lenient in the administration of the Code of Ethics," the author, Secretary of the American Society of Civil Engineers, claims no other engineering society in the world can match its record in the development and maintenance of ethical standards. He then presents a summary of professional conduct cases.

379. Wisely, W. H.: Product endorsement by engineers. *Civil Engineering, 35*:41, April, 1965.

He charges that the American Society of Civil Engineers discusses details and technicalities of the ethical admonition against "personal 'testimonial' advertisements" by engineers.

380. Wisely, W. H.: Spirit of service. *Civil Engineering, 37*:33, December, 1967.

The author, Secretary of the American Society of Civil Engineers, says a profession is characterized and defined by the ethical spirit of service found within it.

381. Wittenberg, Alexander: Ethical issues. *Science, 137*:468-469, August 10, 1962.

In a rejoinder to J. L. Haybittle concerning Haybittle's remarks (*Science, 136*:917-919, June 8, 1962) about the authors call to condemn Russian scientists working on atomic testing (*Science,*

135:997-998, March 16, 1962) the author states that the problem of resolving ethical judgments should not be left entirely in the hands of the individual scientist but should be handled by the professional organizations of science.

382. Wittenberg, Alexander: Standards of ethical conduct. *Science, 135*:997, March 16, 1962.

The author argues that scientists do have a moral responsibility for the use to which their discoveries are put and suggests that professional scientists refuse to recognize colleagues, those Russian scientists who work on atomic bomb testing.

383. Wolfe, Dael: Psychological testing and the invasion of privacy. *Science, 150(No. 3705)*:1773, December 31, 1965.

The author outlines some problems of ethics involved in psychological testing without taking a definite position on any particular solution.

384. Wolfensberger, Wolf: Ethical issues in research with human subjects. *Science, 155(No. 3758)*:47-51, January 6, 1967.

The author discusses the need for a code of ethics in medical research, social science, and behavioral science. He considers, in some detail, the natures of "consent," "research," "risk," and concludes by offering eleven "guidelines."

385. Wolfensberger, Wolf: Letters—'experimentation: rights and risks' (reply). *Science, 155(No. 3770)*:1618, March 31, 1967.

The author answers three replies to his "Ethical Issues in Research with Human Subjects," (*Science*, January 6, 1967).

386. Wolstenhome, G. (Editor): *Man and His Future.* CIBA Foundation Symposium, New York, 1962.

This book contains articles by, among others: Sir Julian Huxley, Colin Clark, Alan Parkes, Donald M. MacKay, Hermann J. Muller, Joshua Lederberg, and more. Many of the articles deal with questions of ethics and science.

387. Woodcock, F. J.: Letters to the editor—the engineer's social responsibility. *Electronics and Power, XIV*:39, January, 1968.

In reply to M. W. Thring's "Social Responsibility of the Engineer" (August, 1967), the author suggests that engineers, who already accept responsibility for the high industrial accident rate, exercise their social responsibility by becoming more concerned about the situation and developing preventive action.

388. Woolhiser, David A., and Falkson, L. M.: Some ethical implications of professional planning. *American Society of Civil Engineers, Journal of Professional Practice, 91(No. 1)*:58-59, January, 1965.

The author's reply to J. M. Abernathy's article of the above title in this Journal, May, 1964.

389. World Medical Association: *British Medical Journal, II*:177, 1964.

"Declaration of Helsinki": a statement on human experimentation. This code of ethics, adopted by the World Medical Association in June, 1964, covers basic principles, clinical research com-
bined with professional care, and nontherapeutic clinical research.

390. Wright, P. M.: Worlds of two giants. *Proceedings of the American
Society of Civil Engineers, 92(No. 2)*:1-5, December, 1966.

The author emphasizes the importance of engineering ethics
in construction and public relations of a contracting firm, in addition to ethics in engineering design.

391. Yale University and WTIC: Ethics and medicine: confidentiality. *Yale
Reports,* No. 467, March 10, 1968.

This report is a transcript of a radio discussion between Dr.
Gerald Klerman, Yale Psychiatrist and the Director of the Connecticut Mental Health Center; Dr. Wilfred Bloomberg, psychiatrist
and Commissioner of Mental Health for the state of Connecticut,
and Catherine Rocaback, attorney.

392. Yale University and WTIC: Ethics and medicine: consent and human
experimentation. *Yale Reports,* No. 465, February 25, 1968.

This report (the first of three on medical ethics) is a transcript
of a radio discussion between Dr. Jay Katz, Yale Professor of Law
and Psychiatry; Dr. Morton Kligerman, Professor of Radiology,
and Guido Calabresi, Professor of Law. *Yale Reports* are available
free from 1773 Yale Station, New Haven, Connecticut 06520.

393. Yale University and WTIC: Ethics and medicine: control of human
experimentation. *Yale Reports,* No. 466, March 3, 1968.

This report (the second of three on medical ethics) has the
same panel as #465. As in the first program (#465), "the concern
is for experiments where the subjects are human beings and the
purpose is to benefit not only the immediate state of the healing
arts, but future patients and the future of medical treatment."

SUPPLEMENTARY BIBLIOGRAPHY

S-1. AIA clarifies its ethical standards. *Engineering News-Record, 173*:12, July 30, 1964.

S-2. AIA to clarify employment ethics. *Engineering News-Record, 173*:23, July 16, 1964.

S-3. Alger, Phillip: The relation of ethics to human progress. *Scientific Monthly*, August, 1942.

S-4. Alger, Phillip; Christiansen, N. A., and Olmstrad, Sterling P.: *Ethical Problems in Engineering.* John Wiley, New York, 1966.

S-5. American Psychological Association: *Casebook on Ethical Standards of Psychologists.* A.P.A., New York, 1967.

S-6. American Psychological Association: Social influences on the standards of psychologists. *American Psychologist, 19*:167-173, 1964.

S-7. Architects move to enforce new ethics. *Engineering News-Record, 174*:22-23, January 7, 1965.

S-8. Architects toughen ethical standards. *Engineering News-Record, 172*: 17-18, June 25, 1964.

S-9. Auger, P.: *Current Trends in Scientific Research.* Paris, UNESCO, 1961.

S-10. Bennett, C. C.: What price privacy. *American Psychologist, 22*:371-376, 1967.

S-11. Berkley, Carl: Opportunities in medical engineering. *American Journal of Medical Electronics, 22*:109-111, 2nd Quarter, 1963.

S-12. Berkley, Carl: Technologic progress and the Hippocratic Oath. *American Journal of Medical Electronics, 3*:1-2, 1st Quarter, 1966.

S-13. Biedermann, H. G.: Ethics of negotiating with prospective employers. *Proceedings of the American Society of Chemical Engineers, 93*: 27-31, December, 1967.

S-14. Boyka, H. (Editor): *Science and the Future of Mankind.* The Hague, Netherlands, W. Junk, 1961.

S-15. Brain, Russell: *Science and Man.* Faber and Faber Ltd., 1966.

S-16. Bronowski, Jacob: *Science and Human-Values.* Harper Press, 1965.

S-17. Budrys, A. J.: *Mind Control Is Good, Bad. Esquire, 65*:106-109, May, 1966.

S-18. Cannons of ethics for engineers. *Ashrae Journal, 4*:18, October, 1962.

S-19. Code of ethics recommended, statement of recommended ethical conduct for the construction industry. *Electrical Construction and Maintenance, 57*:227-229, November, 1959.

S-20. Conrad, H. S.: Clearance of questionnaires with respect to "invasion of privacy," public sensitivities, ethical standards, etc., *American Psychologist*, 22:356-359, 1967.

S-21. Creegan, Robert F.: Concerning professional ethics. *American Psychologist*, 13:272-275, May, 1958.

S-22. Cutler, C. C.: Duty to dissent. *IEEE Spectrum*, 4:47, June, 1967.

S-23. *Daedalus*, Spring 1969—reference whole issue.

S-24. Ethics of planned obsolescence. *Machine Design*, 36:136-138, August 27, 1964.

S-25. Feigl, H., and Maxwell, G. (Editors): *Current Issues in the Philosophy of Science*. New York, Holt, Rinehart, Winston, 1961.

S-26. Fink, B. Raymond: Patient consent. *Anesthesiology*, 28 (6-Pt. 1): 1109-1110, 1967.

S-27. Fletcher, Joseph: *Morals and Medicine*. Beacon Press, 1960.

S-28. Florkin, M.: Medical experiments on man. *UNESCO Courier*, 21: 20-23, March, 1968.

S-29. Fruend, P. A.: Is the law ready for human experimentation. *American Psychologist*, 22:394-399, 1967.

S-30. Gilpin, R.: *American Scientists and Nuclear Weapons Policy*. Princeton, Princeton University Press, 1962.

S-31. Glass, Bentley: *Science and Ethical Values*. Chapel Hill, University of North Carolina Press, 1965.

S-32. Hersh, Seymour M.: *Chemical and Biological Warfare*. New York, Bobbs-Merril, 1968.

S-33. Huxley, Aldous: *Brave New World*. Chatto and Windus Ltd, 1966.

S-34. Ingle, Dwight J.: Biological intervention in human life: the moral issues. *The University of Chicago Magazine*, LXI (No. 2):18-22, September/October, 1968.

S-35. Jaspers, K.: *Future of Mankind*. Chicago, University of Chicago Press, 1961.

S-36. Katz, M. M.: Ethical issues in the use of human subjects in psycopharmacologic research. *American Psychologist*, 22:360-363, 1967.

S-37. Lindsay, R. B.: Physics, ethics, and the thermodynamic imperative. In Baumrin, Bernard (Editor): *Philosophy of Science*, 2:1962-1963. New York, John Wiley and Sons, Inc., pp. 411-448, 1963.

S-38. Lovell, V. R.: The human use of personality tests: a dissenting view. *American Psychologist*, 22:383-393, 1967.

S-39. Masters, William H., and Johnson, Virginia E.: *Human Sexual Response*. Boston, Little, Brown, and Co., 1966.

S-40. Merrie, D. K.: Is the man who designs responsible to society? *Product Engineering*, 38:116-122, December 4, 1967.

S-41. Miller, D. L.: *Modern Science and Human Freedom*. Austin, University of Texas Press, 1959.

S-42. Minker, R. L.: Another look at engineering ethics. *Chemical Engineering, 74*:258+, October 9, 1967.

S-43. Mohan, R. P.: *Technology and Christian Culture.* Washington, Catholic University of America Press, 1960.

S-44. McGehee, W.: And Esau was an hairy man. *American Psychologist, 19*:709-804, 1964.

S-45. NSPE irked by AIA ethical standard. *Engineering News-Record, 173*:29, July 9, 1964.

S-46. Needed: ecumenical ethics: talks at ASCE's Detroit meeting. *Engineering News-Record, 169*:22-23, October 18, 1962.

S-47. Nunnally, J., and Kittros, J. M.: Public attitudes toward mental health professions. *American Psychologist, 13*:589-594, 1958.

S-48. Orwell, George: *1984.* Monarch Press, 1966.

S-49. Parkes, A. J.: *Sex, Science, and Society.* Oriel Press, 1966.

S-50. Pines, M.: Hospital: enter at your own risk. *McCalls, 95*:79+, May, 1968.

S-51. Randal, J.: Merchant doctors. *Reporter, 36*:29-30, May 4, 1967.

S-52. Reagan, Charles E.: Organ transplants and contemporary ethics. *Kansas Alumni,* University of Kansas, October, 1969. Reprinted in *Kansas City Star,* November 6, 1969.

S-53. Revised and expanded code of ethics put to members; ASCE meeting, Phoenix, Arizona. *Engineering News-Record, 166*:24-25, April 20, 1961.

S-54. Rostand, Jean: *Can Man Be Modified?* New York, Basic Books, Inc., 1959.

S-55. Rychlak, J. F.: Control and prediction and the clinician. *American Psychologist, 19*:186-190, 1964.

S-56. Schuster, R.: Profile of a real professional. *Plant Engineering, 21*:127-130, April, 1967. (*22*:97-102, April 18, 1968.)

S-57. Schwitzgebel, R.: Electronic innovation in the behavioral sciences: a call to responsibility. *American Psychologist, 22*:364-370, 1967.

S-58. Seaborg, G. T.: *Freedom and the Scientific Society.* Williamsburg, Virginia, Colonial Williamsburg, 1962.

S-59. Shideler, M. M.: Coup de grâce." *Christian Century, 84*:20, 82-83; 272-273, 471; January 4, 18, March 1, April 12, 1967.

S-60. Shumway, N. E.: State of many arts. *Science News, 93*:213-214, March 2, 1968.

S-61. Smith, M. Brewster: Conflicting values affecting behavioral research with children. *American Psychologist, 22*:377-382, 1967.

S-62. Standards of ethical behavior for psychologists. Report of the Committee on Ethical Standards of Psychologists. *American Psychologist, 13*:266-271, May, 1958.

S-63. Stevens, H. A. In Oster, J. (Ed.): *International Copenhagen Congressional Science, Study of Mental Retardation.* (Copenhagen, Berlingske Bogtrykkeri, 1964) Vol. 1, p. 37.

S-64. Surgeon General's direction on human experimentation. *American Psychologist, 22*:350-355, 1967.

S-65. Thring, M. W.: Social responsibility of the engineer. *Electronics and Power, 13*:292-294, August, 1967.

S-66. Turner, E. M.: Human side of engineering. *IEEE Spectrum, 4*:70-71, July, 1967.

S-67. Waddington, C. H.: *Ethical Animal.* London, Allen and Unwin, 1960.

S-68. Watson, Goodwin: Moral issues in psychotherapy. *American Psychologist, 13*:574-576, October, 1958.